KILLING OUR OCEANS

Dealing with the Mass Extinction of Marine Life

JOHN CHARLES KUNICH

Westport, Connecticut
London

Library of Congress Cataloging-in-Publication Data

Kunich, John C., 1953–
    Killing our oceans : dealing with the mass extinction of marine life /
John Charles Kunich.
        p.   cm.
    Includes bibliographical references and index.
    ISBN 0–275–98878–3 (alk. paper)
    1. Marine biological diversity conservation—Law and legislation.   I. Title.
K3488.K86   2006
346.04'695616—dc22   2006004355

British Library Cataloguing in Publication Data is available.

Library of Congress Catalog Card Number: 2006004355
ISBN: 0–275–98878–3

First published in 2006

Praeger Publishers, 88 Post Road West, Westport, CT 06881
An imprint of Greenwood Publishing Group, Inc.
www.praeger.com

Printed in the United States of America

∞™

The paper used in this book complies with the
Permanent Paper Standard issued by the National
Information Standards Organization (Z39.48–1984).

10 9 8 7 6 5 4 3 2 1

CONTENTS

PREFACE

"We have it in our power to begin the world over again."[1] Those were
the words of Thomas Paine, one of the founders of the United States of
America, in 1776. He was referring to the world in the political sense,
and the possibility that a group of brave, determined, visionary people
could break free from the injustices that bound them, and together
create a new environment based on liberty and human rights. Events
proved Paine's famous words to be prophetic. But today a different set
of wrongs demands bold and courageous action, and this time the
"world" that requires a new beginning is the physical world itself.
Planet Earth, and countless living things on it, are threatened by the
shortsighted and misguided actions of people—including many in the
United States—and unless enough people soon resolve to set a new
course, the consequences will be both devastating and irreversible. We
often hear people say, "That means the world to me." But what does
the world itself mean to us? Are we willing to do what it takes to save it?

Killing our oceans? The title of this book sounds like one more
overblown, alarmist attempt to scare people into doing something,
anything—such as buying the book. But the truth hurts, and in this

case it hurts a lot more than just people. A mass extinction now threatens much of life on Earth, and marine life is particularly at risk. We are currently in the midst of at least the sixth mass extinction in this planet's history, one of the catastrophic death spasms in which vast numbers of species disappear forever at far greater than the usual rate.[2] In this book I will examine the appalling extent to which the Sixth Extinction has reached the world's oceans, and I will demonstrate that stacks of international and domestic laws have done nothing more about this devastation than act as a dangerous placebo. My conclusion will provide an antidote to this syndrome of law as the new opiate of the masses which soothes us to sleep, secure in the delusion that life on Earth is safely protected by legions of laws.

Our collective image of life in the seas is still shaped today by the stories and memories from only a few decades ago. Some people alive in 2006 were witnesses to the teeming waters of not many years past, waters bursting with seemingly limitless schools of great fish. Visions of living waves of numberless marine organisms of all sizes and varieties attained near-mythic status in the minds of many people. These pictures of oceans literally overflowing with infinite expanses of vibrant life—swirling, silvery clouds of swimming swarms—remain locked in our common assumptions, and serve to fill the large voids of hard facts about marine biodiversity as it really exists now. As I will show, in far too many cases these epic cascades of hyperliving seas are no longer anything but a fading ripple. Mass extinction has left emptiness in the once-crowded waters.

In three previous articles and a book I have established the dearth of effective legal protection for the planet's terrestrial biodiversity hotspots and the nameless hosts of species crowded in them. On dry land, and in the internal fresh waters of the world's nations, there is no comprehensive, efficacious, enforceable legal mechanism in place—not in terms of U.S. legislation[3]—nor in international law or the aggregate laws of the various nations that are home to the hotspots.[4] Even the best of international treaties have failed to make a discernible dent in the dreadful loss of key habitats.[5] This is a disastrous state of affairs because the hotspots are the sole repositories of an immense share of all remaining life on Earth. If they are lost, countless species will vanish with them.

In this book I will move the hotspots focus to a very different, yet in some ways quite similar, aspect of our contemporary mass extinction. If there are titanic unknowns riddling the question of terrestrial extinctions—and there are—the situation is even more extreme when we move from our comfortable and familiar land-based environment and venture into the oceans.[6] There, amidst the vastness and darkness, we know virtually nothing about the most vital and most ancient of habitats.

I will set forth some essential background information as to what little we know about marine biodiversity, how many species exist in the world's oceans, where those species are concentrated, and what threats challenge their continued existence. Then I will examine the current legal protections that theoretically stand in opposition to a marine mass extinction, yet have been powerless to prevent or arrest the wholesale emptying of the waters. Finally, I will conclude with a paradigm-shifting proposal for a more effective legal approach to safeguarding Earth's marine life. I will make every effort to stay away from fancy jargon, difficult and painful as that is for someone who is a law professor by profession. My goal, above all, in this book is to educate and to persuade people that something of incredible value is being irretrievably lost, right now, right below the waves, and we need to take swift action to prevent it.

ACKNOWLEDGMENTS

I thank my wife, Marcia Kathleen Vigil, and our daughters, Christina Laurel Kunich and JulieKate Marva Kunich, for their wonderful love and support, without which this book would not have been possible. I also gratefully recognize the superb research contributions of John Harrington and Deborah Niedfeldt to this book.

ONE

Hotspots Under the Sea: Hotter Under the Water?

MASS EXTINCTION BY THE NUMBERS

With regard to extinction spasms, Earth's oceans, along with all other habitats, have been there, done that, long before now. It is generally accepted that there have been no fewer than five mass extinctions in Earth's history, at least during the Phanerozoic Eon (the vast expanse of time which includes the present day). These "big five" mass extinctions occurred at the boundaries between the following geological periods: Ordovician-Silurian (O-S); near the end of the Upper Devonian (D) (usually known as the Frasnian-Famennian events, or F-F); Permian-Triassic (P-Tr); Triassic-Jurassic (Tr-J); and Cretaceous-Tertiary (K-T).[1] In terms of millions of years ago (Mya), the mass extinctions have been placed at roughly 440 for O-S, 365 for F-F, 245 for P-Tr, 210 for Tr-J, and 65 for K-T,[2] with the mass extinctions taking place over a span of time ranging from less than 0.5 million years to as long as 11 million years.[3] There is some evidentiary support for other mass or near-mass extinctions in addition to the big five, including events near the end of the Early Cambrian (about 512 Mya) and at the end of the Jurassic and Early Cretaceous, among several others.[4]

I will return to this later, but it is important to note at the outset that these ancient mass extinctions, devastating as they were, most emphatically did *not* happen with anything approaching our modern notion of swiftness. Irrespective of the primary cause, each of these extinction spasms unfolded with what would seem to us as imperceptible gradualness, the "spasms" lasting hundreds of thousands, and even millions of years. For anyone who has ever suffered through a back spasm or a charley horse, imagine enduring such agony for millions of years and you will have some grasp of the horrific and prolonged nature of an extinction spasm. In fact, "mass extinction" is a relative term, because there is always at least a background rate of extinction as species naturally live out their life span and go out of existence. There is nothing unusual or catastrophic about an occasional extinction; it has been happening for as long as there has been life on Earth, and for hundreds of millions of years before people existed. Mass extinctions are simply periods of time in which there is an extinction rate far greater than the norm, although still very slow from the highly limited perspective of human day-to-day time standards. The normal background pace of extinction is significantly accelerated during a mass extinction, so that many more species than usual cease to exist per unit of time—but to people brought up on MTV and *Sesame Street*, it still appears that nothing out of the ordinary is happening, and that is a large part of the problem.[5] It is calamity masquerading as calm.

Although much has been written in scientific literature about these historical mass extinctions, relatively little attention has been devoted to extinctions in the oceans.[6] And especially for those marine areas that generally remain submerged under thousands of feet of sea water, the usually formidable challenges of piecing together the ancient evidence are greatly magnified. This is the ultimate example of a "cold case," literally and figuratively, because the evidence of these long-ago events is so hard for us to reach, covered as it is by water of prohibitive pressure and frigid temperatures, all in total darkness.

It is extremely difficult to arrive at a satisfactory estimate of the magnitude of the current extinction crisis, whether in the marine realm or on dry land. One problem we face is that we know so little about life on Earth today in the first place, even in areas much more accessible than the oceans' depths. If we do not know how many species exist, we cannot know precisely how many are ceasing to exist; respectable estimates as to the number of species now in existence vary by an order of magnitude (i.e., a factor of ten). Moreover, for many of the species we have identified, we know very little about their range, their habits, their life cycles, and other details important to an understanding of their health or risk status.

Although there is some scientific dispute due to the gaping gaps in our information base, by far the most widely held expert view is that the Earth is now in the midst of a sixth mass extinction that rivals the great disappearances of ages past.[7] The overwhelming weight of the evidence points unmistakably to the conclusion that the vast majority of species now alive will be extinct long before scientists have even identified and named them. This is not some crackpot theory cooked up by a mob of howling zealots who want to return the world to some mythical preindustrial Shire-like utopia. I wish it were. But this is not one of those topics about which there is a good, old-fashioned scientific brawl raging. If there is a debate, it is hard to hear any respectable scientists arguing for the other side. There is the closest thing to a scientific consensus you will ever find (though you would not know it to read the newspapers or watch the news on television) that we are living in a world in the death throes of a mass extinction the likes of which this planet has not seen in 65 million years.

In part, this decimation of life on Earth is being caused by direct killing, usually overhunting and overfishing; it is also being caused indirectly by the introduction of exotic or invasive species into new habitats where they outcompete the native species. As I will explain shortly, both of these issues have severe effects on living things, both on land and in the oceans, and the impacts are powerfully felt far beyond the species and the zones that are most directly and immediately hit. But the greatest destructive force is as slow-acting as it is deadly: the harmful modification and obliteration of the most vital habitats of the world.

It makes sense that this would be the number one cause of death for species. Destroy the home, and you destroy the inhabitants. This is especially true for the many species that are narrowly adapted to live only within a specific set of conditions of temperature, salinity, light, terrain, food supply, and other factors. Specialization for life under such a well-defined set of circumstances can be a very successful strategy for species survival, but only if those circumstances are not disturbed. If they are, then all bets are off, and all those millions of years of adaptation to that one set of conditions spells out a death sentence. And, sad to say, habitat destruction is one of the things people do best. We have plenty of experience.

There were those who foresaw this devastation coming over the horizon, long before it reached the present emergency level, but as usual the warnings were mostly dismissed. For example, in his seminal work on the extinction situation decades ago, renowned British ecologist Norman Myers of Oxford predicted the current extinction crisis, primarily a result of habitat destruction and other human actions.[8] Myers warned that the world could soon suffer an "extinction spasm accounting for 1 million species." Tragically, his estimates may have been overly optimistic, as he himself now recognizes.[9] To put this in historical context, the background or natural rate of extinction has been estimated to average only a few species lost per million years for most taxonomic groups.[10]

Predictions and diagnoses of a contemporary mass extinction are generally derived by extrapolation. Larger, more well-known species, such as mammals and birds, are more visible, more easily studied, and much more thoroughly identified and catalogued than most aquatic life forms and invertebrates. Mammals and birds also are well represented in the fossil record, enabling scientists to craft better estimates of their historical extinction rates than groups that do not lend themselves as well to fossilization.[11] Thus, mammals and birds can be used as indicators or proxies for other groups' extinction rates and histories because they are (1) taxonomically relatively well known,[12] (2) easily observed, and (3) prominently etched in the fossil history.

Mammals and birds, however, constitute only a small minority of the community of living things, both in terms of number of species and in terms of number of individuals. Invertebrates, particularly members of the phylum Arthropoda and, within it, the class Insecta, account for the vast majority of described species. More than one million species of insects have been given scientific names.[13] Enormous as this total is, though, some have opined that this may amount to less than 10 percent of insect species, with particularly large numbers of unknown species presumed to reside in the tropics.[14]

It has been estimated that the ratio of unknown to known species may be as high as 21 to 1, with 30 million undescribed species versus the approximately 1.75 million that have been identified and taxonomically categorized by people.[15] Some biologists hold that the great majority of the species of insects, nematode worms, and fungi have yet to be discovered.[16] Although no one knows for certain, there seems to be an emerging scientific consensus that the total number of species on Earth today is somewhere in the range of 7 to 13 million, with the best "rule of thumb" estimates centering on 10 million species, very roughly speaking.[17]

This is one of those subjects on which you could venture almost any semieducated guess and no one could definitively prove you wrong. After all, how can you know how many species you do not know about? It would be like asking a person to guess how many opportunities for a better career she has missed because she went into the wrong line of work. But expert scientific experience with tropical insects, deep sea microorganisms, and other rather humble and remote life forms (for which you only need to do a little looking in the right places to discover species new to science) has led to these widely accepted guesstimates of millions of unknown species alive today. Indeed, if it was possible for three disorganized, work-averse, beer-drinking graduate students (of which I was one) to discover new species of flies during three weeks of utterly sporadic insect collecting in Peru in 1978, there must be a lot of unknown species out there, just waiting for someone to find them.

When attempting to gauge the dimensions of the current mass extinction, there are some other knotty obstacles besides not knowing how many species we have to begin with. For one thing, it is very difficult to determine when the last individual member of a species has died, a feat akin to proving a negative. It is virtually impossible to monitor the fate of many small and obscure species, particularly when the species only exist within a remote, inaccessible wilderness or a deep, dark ocean abyss.[18] There can also be a very long lag time between the point at which a species becomes "committed to extinction" and the point at which it actually becomes extinct (a key point relevant to the ultimate effects of rampant habitat destruction).[19] In other words, a species may be given a death sentence by such pressures as severe habitat constriction, thereby being condemned long before its ultimate extinction, and spend many years languishing on "death row." The "living dead" in this situation are not yet extinct but are irreversibly on the way there, a nonhuman variation of the "dead man walking" theme.

Simply put, a species becomes committed to extinction when one or more factors combine to make it impossible for it to recover enough to survive indefinitely. If it needs a specific set of conditions to live, and the available habitat where all those essentials are met has drastically shrunk, there can be inadequate space, food, and living niches to support the species in sustainable numbers. For every species there is some minimum number of reproductively capable individuals necessary to provide sufficient genetic diversity for vigorous offspring and the ability for the population as a whole to withstand onslaughts from disease, predation, adverse weather, famine, hunting, natural disasters, and competition. If there are too few individuals, there will not be enough of a cushion, not enough of a margin for error, to see the species through the inevitable, if rare, periods of crisis. It may be millennia before a given "living dead" species is unlucky enough to come face to face with the ultimate catastrophe that tips it over the tipping point into actual extinction, but it is just a matter of time.

Of course, many factors coalesce to determine the point at which any particular species crosses that fateful threshold and becomes committed to extinction, and no one knows exactly what that point is. It varies from species to species, and changes over time as conditions evolve and new or different threats develop. But it is not too hard to understand the concept, once you think about it. The real challenge is to gauge how many species we might now number among the living dead, and what human actions are pushing these and other species toward that lethal cliff.

This phenomenon, in which rampant habitat destruction forces many species into a situation of inevitable but not immediate extinction, is especially insidious. Because the ultimate extinction of most species is postponed, perhaps for a great many years, people may be misled that all is well. We can still see for ourselves real-life examples of some of these species, often in zoos or wildlife refuges, and they certainly do not look extinct when we see them alive and kicking, or flying, or swimming. These survivors may be few in number, and we may only see them alone or in pairs, but they just do not seem extinct to us when the evidence right before our eyes tells us that some of them are still alive. But these species that are doomed to die—the "living dead"—*will* become extinct in due course, turning into the "no-kidding" dead, and by the time the actual extinctions commence in large numbers, it will be too late. Like an immense balloon payment on a mortgage that only comes due at the bitter end of a very long period of living on borrowed time, we are on the road to an avalanche of dying species.

The unknown but still living species pose another challenge to anyone trying to assess ongoing mass extinction. If we do not know that a species exists, can we know when it ceases to exist? It makes sense, though, that extinction risk would generally be *more severe* among the unknown species than among the known. This is true, in part, because there may simply be many more species without a name than with one (as plenty of experts believe), so there is statistically a greater chance for them to go extinct. But also, the reason we have not identified and named these species may, in many cases, be linked to their rarity. In numerous instances, there probably are not enough of them in enough places for us to have a reasonable chance of discovering them. The members of unknown species might very well tend to be less abundant and less widely distributed, and thus more prone to extinction than species with larger populations and a foothold in lots of different habitats.[20] In fact, it would be very surprising if this were not the case, because there is likely a good reason why no one has ever described all these unnamed species, and that reason often has to do with their rarity and their very limited distribution. If the members of these unknown species were plentiful and widely dispersed, it is probable that someone would have discovered them already. And what is a better predictor of high extinction risk for a species than low numbers and extremely narrow habitat requirements?

At this point, a brief overview of some other matters is in order if we are to understand the biodiversity crisis in the world's oceans. Let us examine the ways in which we categorize these waters. I will begin with categories based on natural features, such as depth of water and type of inhabitants, followed by a discussion of the threats to marine biodiversity.

ZONING IN THE OCEANS

Broadly speaking, the Earth's oceans are divided into the pelagic zone and the benthic zone, both of which are astonishingly enormous when compared to any terrestrial region. One can gain some appreciation for this by considering the fact that the pelagic zone essentially consists of the entirety of the actual waters (or water column) in all the oceans on the planet. Equally vast is the benthic zone, which is the ocean bottom or floor, as opposed to the water column above it.

The pelagic zone is further subdivided into three zones on the basis of depth. Nearest to the surface of the water is the euphotic zone, also known as the sunlight or sunlit zone, which consists of the waters through which significant amounts of sunlight can penetrate. The euphotic zone extends from the ocean surface down to a depth of about 660 feet. This area is of great ecological importance and richness because the sunlight enables photosynthesis to occur. The resulting profusion of aquatic plant life serves as a bountiful and varied source of nutrients for a host of life forms. The water temperature is also relatively warm here, again owing to the presence of sunlight and the frequent mixing of water that takes place, and this hospitable warmth contributes to the proliferation of species.

Immediately beneath the euphotic is the dysphotic zone, which is also called, eerily enough, the twilight zone. Although some sunlight still penetrates this region, it is not in sufficient quantities for photosynthesis. Thus, the dysphotic zone is virtually devoid of photosynthetic plants, with a corresponding diminution of other living things. The dysphotic zone is generally considered to stretch from about 660 feet to about 3,300 feet below the surface. This region is naturally both darker and colder than the euphotic zone, and on the whole these conditions have traditionally been assumed to be less conducive to a multiplicity of species, although that may not be the case.

The next stop on the way down after the twilight zone is the aphotic or midnight zone. There is a total absence of sunlight in the aphotic zone, which can stretch from approximately 3,300 feet to about 20,000 feet beneath the surface. Sometimes an additional subdivision of the aphotic zone is recognized—the abyss—consisting, as you might expect, of the very deepest waters. Some areas of the abyss, such as the Mariana Trench, are so deep that Mount Everest (all 29,035 feet of it) could be submerged in them, with plenty of room to spare.[21] Of course, no photosynthesis is possible in any part of the aphotic zone. These profoundly dark, frigid waters are prohibitively severe for many types of life, especially given that the extremes of dark and cold are exacerbated by crushing water pressure. Still, there are many other living things even at these incredible depths, and we have only begun to explore the highly specialized biodiversity that exists here. In fact, because colder water has a greater capacity to hold gasses, including oxygen and carbon dioxide, even the most frigid water can support a large population of living things, as exemplified by the freezing waters of the Arctic and Antarctic.

To muddy the waters a bit (pardon the pun), there is another type of nomenclature for ocean zonation based on depth of the water column. In this system, the neritic zone consists of the portion of the pelagic zone that extends from the high-tide line to an ocean floor less than 600 feet below the surface. The remainder of the pelagic zone (i.e., water of a depth in excess of 600 feet) is called the oceanic or open ocean zone, which in turn is divided into the epipelagic, mesopelagic, and bathypelagic zones, based on the amount of sunlight that penetrates.[22] Roughly speaking, the epipelagic zone corresponds to the euphotic or sunlight zone; the mesopelagic corresponds to the dysphotic or twilight zone; and the bathypelagic corresponds to the aphotic or midnight zone, where by far the greatest share of the water column is found.

But wait, there's more. The ocean is also sometimes divided into two main segments. The first of these is the continental margin, which is composed of the continental shelf and slope (15.3 percent of the total ocean) and the continental rise and sedimentary basins (5.3 percent of the total ocean). The second is the deep ocean, which is composed of the abyssal plain (41.8 percent of the total ocean), oceanic ridges (32.7 percent of the total ocean), and other areas (4.9 percent of the total ocean).[23] The depths of these areas vary according to the unique physical characteristics of a particular area. In general, the continental slope ends at a depth of about 4,900 and 9,800 feet and the continental rise meets up with the abyssal plain at about 11,400 to 16,400 feet.[24]

WHAT ARE HOTSPOTS?

I have written elsewhere about our monumental ignorance of the biodiversity in Earth's terrestrial hotspots.[25] The *hotspots*—those relatively few habitats that for various reasons are the only home to far more than their fair share of living things—are extremely important, because if they are lost, they take all the life they contain with them. We know that these key habitats are the last remaining sanctuaries for hundreds of thousands of identified species, and that alone would be ample reason to preserve them; but their significance extends far beyond that. As I mentioned earlier, the great weight of respectable scientific evidence is in support of the proposition that millions of unnamed species, completely unknown to humanity, inhabit these hotspots. If we have not even assigned a name to these numberless, nameless species, we certainly have no inkling as to their ecological significance or their potential utilitarian value for human beings. This is not a case of a pig in a poke; it is more akin to millions of whatchamacallits in the smoke and shadows. And just as there are certain limited habitats on land that provide the only abode for a disproportionate number of species, there are also key areas in our oceans that are ideal for many species found there *and nowhere else.*

Some marine areas are notably different from the "typical" (if there is such a thing) ocean habitat. By virtue of their proximity to the surface, availability of major currents, amount of and fluctuations in warmth, pressure, degree of penetrating sunlight, abundance of complex substrate with niches and nooks (not to mention crannies) to live in, nearness to land, ready availability of various types of food, protection from violent storms, and an array of other features, these places are able to offer a suitable or even ideal home to many creatures. For species with unusual and highly specific needs, certain combinations of factors are absolutely essential for their survival, and they *must* live in these super-habitats. There is no other option for certain creatures.

The oceans are not uniform, homogeneous, and fungible. To the uneducated eye taking a fleeting glance seaward, water might look like water anywhere in the world. However, this superficial and simplistic view is the opposite of the reality. There are vast differences within the oceans, from place to place, and some areas are immeasurably more hospitable to most forms of life than others. These vital habitats are the marine hotspots. They are the *only* home for myriad living things that do not and cannot exist anywhere else. If the hotspots are mined, polluted, overfished, dredged, or otherwise altered, the results will, with logical inevitability, be disastrous. But where are these ocean hotspots, how many species live in them, and what is the magnitude of the risks they face?

If there are so many unknowns about life on good old dry land (and there are), then those unknowns must be adorned with an exponent when the habitat in question becomes the much less human-friendly oceanic realm. This is quite understandable, given that people are terrestrial, air-breathing, freshwater-drinking, nonmarine mammals. We are very much out of our element in ocean water. Even in shallow salt water near shore, we require scuba gear to perform more than the most cursory examination of aquatic life forms, and then we are limited to brief forays. As we move into deeper water, we rapidly lose the reassuring presence of the sun's light and become dependent on artificial sources of illumination as well as air. The water also becomes uncomfortably, even perilously cold, and we need special suits to stave off deadly hypothermia. The water pressure grows so great that our fragile bodies soon become unable to withstand the crushing, pythonlike force exerted upon us.

In general, the depth of the ocean's water is positively correlated with distance from shore, which adds further obstacles to human exploration. As we travel farther from our land base, we need larger and more rugged vessels; swimming alone is no longer sufficient to get us there and back again alive. The greater the distance from shore, the longer the voyage must be, and this necessitates more supplies, more fuel, and more money. To delve into any waters but the uppermost environs of the euphotic zone, we must usually resort to some type of submarine or submersible capsule in order to withstand the excessive pressure. This generates more technological challenges, and more expense. In fact, for some voyages into the dysphotic and aphotic zones, there may be no vessel in existence that is up to the rigors of the environment. Hard as it may be to imagine, there are many parts of the ocean, right here on Earth, that are as impossible for us puny humans to reach as the rings of Saturn. The old saying "You can't get there from here" is literally true.

Thus, some of the core elements of the hotspots concept—the unknown species, in unknown numbers, of unknown value—are magnified in the oceans. Other than the biota in a few relatively accessible coral reefs, marine life has been so difficult to study that we know very little with any appreciable degree of certitude. We have no idea what is down there. This situation lends new meaning to the word "unfathomable."

THE AMAZING DIVERSITY OF MARINE LIFE

To illustrate the sheer magnitude of biodiversity in our oceans, let me ask you a question. Would you consider it newsworthy if an entirely new kingdom of living things were discovered? Kingdoms, of course, are generally considered the very highest and most expansive taxa recognized, above the levels of phyla, classes, orders, families, genera, and species, and all the associated super- and sub-taxa at various levels. Traditionally, taxonomists had recognized no more than five kingdoms: Animalia, Plantae, Monera (microorganisms without a distinct nucleus, such as bacteria), Fungi, and Protista (microorganisms possessing a distinct nucleus, such as algae, protozoans, and slime molds). But in 1997 a new kingdom of life forms was recognized, at least according to some taxonomists. Kingdom (or, in the opinion of some scientists, Domain) Archaea, the tiny and astoundingly ancient members of which exist today mostly in association with hydrothermal vents, belatedly joined the highest pantheon of living things. Why did it take so long? Because so little is known about the deep ocean that it was not until 1977 that hydrothermal vents were even discovered. At first it was only in these highly extreme conditions that Archaea were thought to exist (although recently they have also been discovered in such unusual habitats as marshlands, sewage treatment plants, and some animal digestive tracts).[26] Archaea may be picky about where they live, but they certainly will not have to compete with humans for these homes.

If you are not a biology nerd like me, you may not get too excited about the possible arrival of a new kingdom on the scene. Indeed, many laypeople probably still believe that there are only two kingdoms of life: Plant and Animal. But for professional taxonomists (the scientists who devote themselves to studying the degrees of relationship between and among living things), an absolutely huge number of differences must exist between two groups before they can be considered members of separate kingdoms; for example, it takes more differences than exist between a housefly and an elephant, or between an earthworm and a gorilla, or between an oyster and a butterfly, or between a leech and a robin, or between a tuna and a sea star. The vast differences between and among all of these examples have not been deemed sufficient to place any of these creatures in any kingdom other than Animalia. *That* is how significant it is to be classified as a new kingdom. And now we may have an entirely new kingdom, or alternatively, a major new "domain" of life forms (which, in one taxonomic system, is an even higher category than kingdom), consisting of extraordinarily ancient bacterialike entities found mostly in the oceans. This is a powerful testament to the vital importance of marine life.

How many species exist in the world's oceans? This is a question to which no one knows the answer. A person could venture any guess and no one could prove her wrong. In fact, no one knows with any reasonable level of confidence how many species exist on dry land, as I mentioned earlier, and there is far more uncertainty in the ocean's waters. However, within the quite limited universe of the species we have named, we can definitely state that, of the total number of known species, only around 15 percent are marine.[27] Put another way, humans have at present identified approximately 300,000 marine species worldwide out of a total of 1.75 million species.[28] However, credible scientific opinion holds that, as with terrestrial species, only a small percentage of marine species are actually known.[29] Estimates of the total number of ocean species vary greatly.[30] There is no question though that the higher-taxa diversity of the marine environment is much greater than that of the terrestrial environment.[31]

Because phylum-level diversity is far more indicative of great differences in genetic content, evolutionary divergence, form, and function than species-level diversity, it is clear that oceans represent *twice* the diversity of all terrestrial habitats combined.[32] The reason there is so much diversity may be that the oceans are where life on Earth originated (hence the extremely ancient pedigree of the Archaea, for example), and evolution has been going on in oceans for much longer than on dry land. Plus, the oceans have double the surface area and about two orders of magnitude (i.e., 100 times) more biological volume than land, providing far more biogeography.[33] In terms of biogeography, there may be at least 300 different marine biogeographic "provinces" when we include midwater, deep-water, and off-shelf benthic areas, as opposed to the 193 biogeographic provinces identified on land.[34]

What is the significance of the fact that at least fourteen, and as many as twenty-one, phyla of living things are *confined entirely* to the marine environment? Most of us never even think about the concept of the phylum in our everyday lives, so it can be a difficult point to grasp. But consider for a moment that every mammal, every bird, every fish, reptile, and amphibian in the world—all those creatures so familiar to us—belongs to a single phylum: the Chordata. So many species, of such astounding variety, ranging from the mouse to the bald eagle to the great white shark to the sea tortoise to the king cobra—yet all are members of Phylum Chordata. Similarly, think about all the insects you are aware of, from the common cockroach to the most beautiful tropical butterfly. Stir in head lice, mosquitos, the huge Atlas moth, silverfish, fleas, honeybees, dragonflies, and the tiniest ant. Add every spider from the black widow to the giant tarantula, plus all the ticks and mites and other arachnids. Now combine this mixture with crustaceans like lobsters, crabs, crayfish, and shrimp. It is a conglomeration of mind-boggling differences, and you wouldn't want to dive into a pit filled with all of these critters. But again, they all are members of just one phylum: Arthropoda. These two examples serve to illustrate the stunning amount of biodiversity subsumed within every phylum—how extremely different species can be and still be classified within the same phylum. You can begin to see how divergent two species must be to be considered members of different phyla—and how amazingly diverse marine life must be to encompass a minimum of fourteen phyla found solely in the oceans and nowhere else!

People are often surprised to learn of the vast spectrum of species that can belong to the next taxonomic layer beneath phylum—the class. To confine myself to my personal favorite, the class Insecta,[35] I will offer a few disparate examples that illustrate the breadth of this single class of living things. This one class includes: the sunset moth of Madagascar (a day-flying moth with gorgeous iridescent rainbow-hued wings); the pesky housefly; the huge goliath beetle of Africa; the termite; the enormous and brilliantly colorful birdwing butterflies of Papua New Guinea; the much-despised cockroach; the ladybird (often called ladybug) beetle; the primitive springtail; the seventeen-year cicada; the stonefly; the flea; the damselfly; the praying mantis; the giant walking-stick of Asia; the common grasshopper; and many more examples of wondrous variety. This single class contains more than one million named species, with some 350,000 identified species of beetle (the Order Coleoptera) alone. That means there are six times as many known species of beetle as there are of all vertebrates combined. We can see that within any of the three highest taxa—kingdom, phylum, or class—there can be a prodigious wealth of biodiversity. And thus, when we understand that the oceans have much greater higher-taxa diversity than any terrestrial habitat, the paramount importance of the marine environment becomes clear.

This quantum of biodiversity at the very high taxonomic level of the phylum, and even the kingdom, is a shorthand way of expressing the staggering extent of evolutionary adaptation and the hundreds of millions of years of evolutionary history represented by marine life. Because oceans are the only home to so many diverse life forms, separated from one another by so great a portion of Earth's history, our oceans are the most vital repository of living things on the planet. If we wish to find unique genetic codes, or novel adaptations to extreme environmental conditions, or untapped sources of new medicines and nourishment, it is a smart decision to begin our search in the place most likely to hold the answers: the oceans.

There is powerful, if anecdotal, evidence that there are myriad marine species still to be discovered. For example, according to the latest Census of Marine Life, new marine fish species are now being identified at a dizzying rate, with some 600 previously unknown species catalogued since 2000.[36] Marine molluscs are being discovered at a pace of about 300 new species per year.[37] When this explosion of new species is combined with the great preponderance of marine phyla over terrestrial phyla,[38] it becomes clear that it is difficult to overestimate the extent to which the limits of marine biodiversity have yet to be imagined. The latest totals of known marine species are just over 15,300 species of fish[39] and hundreds of thousands of oceanic species overall.[40] How high these numbers will climb during the next couple of decades is a matter rife with conjecture.

What of the old notion that the deeper waters and ocean floors may be comparatively sparsely inhabited?[41] While it was once generally accepted that biodiversity decreased with increased depth,[42] recent studies suggest that this is not true. One study found that the diversity of benthic organisms actually peaks at a depth of between 4,900 and 6,500 feet.[43] The old view of the open ocean as a watery desert has been thrown overboard by new evidence such as the discovery of an enormous, long-overlooked biomass in the pelagic zone, composed of tiny organisms called picoplankton and nanoplankton, which supports tremendous production.[44] Estimates as to the numbers of different species of deep-ocean invertebrates range from 200,000 to as many as 10 million, a staggering degree of biodiversity rivaling or surpassing even that of insects.[45] Could there really be 10 million undescribed species in the deep ocean? There is evidence in support of this astonishing idea.[46] It is symptomatic of our profound ignorance, however, that another study estimated the number of unknown species in the deep ocean at "only" 500,000[47]—a mere half a million species still waiting to be discovered. When scientific specialists arrive at estimates that differ by a factor of twenty, we are dealing with an astonishingly difficult problem.

Perhaps a factoid will help illustrate the prodigious magnitude of the task of exploring life in the oceans. Consider that, out of the 2.9 billion square feet of deep ocean (benthic) floor in the entire world, scientists have sampled only around 5,400 square feet.[48] What would it take to begin to rectify this towering lack of knowledge? If we wished to sample merely one one-millionth of the ocean floor, we would need to study 5,400 square feet a day for a full millennium.[49] You read that correctly. We would have to plunge in and work furiously, and every day we would have to match the sum total of all the benthic zone territory ever explored in the history of humankind. Yet even if we did this every single day for the next 1,000 years, we still would have seen only one part out of one million of what there is to see. It is small wonder, then, that we know more about the fourth rock from the sun, our distant neighbor Mars, than we do about life in the oceans of our own home planet.

ENDEMISM: WHAT IS IT AND WHY DOES IT MATTER?

Some evidence indicates that the majority of marine species, both discovered and undiscovered, exist in the deep ocean benthic mud, the most inaccessible habitat on Earth. Additionally, there are other unique marine environments, such as seamounts and hydrothermal vents, which deserve special mention. Such extraordinary habitats are natural candidates for any list of marine hotspots, precisely because of the distinctive, and even unparalleled, compliment of characteristics they possess. Their degree of difference from the norm often means that they will exhibit a high rate of endemism—certain species will tend to be found only in the habitat these hotspots offer.

Endemism is a concept unfamiliar to many people, but it really is not difficult to grasp. Broadly speaking, there are two main strategies employed by living things as they strive to survive. One strategy of survival is to become as flexible and adaptable as possible to a wide spectrum of food, temperature, weather, predation, and other conditions. The living things that are able to follow this path will, quite naturally, tend to be widely distributed across a variety of different habitat types. Because they are so ubiquitous and adaptable, they will be very resistant to extinction. Even if some portions of their range are badly damaged or destroyed, or there is a significant change in climate, these species will usually be able to weather the storm, so to speak. But the other primary strategy is very different. Many species evolve so as to be extremely well suited to a rather precise set of conditions, as specific as the other group's was general. They develop a highly effective and specialized means of exploiting a particular type of food source, climate, or type of terrain, and they become masters of their limited domain. This strategy can be wonderfully successful for many millions of years, and will remain so indefinitely . . . but only so long as key portions of the habitat stay within the narrow boundaries required by the species.

Species that adopt the precision, specialist approach to adaptation have a much more limited range than their generalist relatives. They can live only where conditions are suitable to their, shall we say, picky and finicky lifestyle. Where they have just the right food, climate, and shelter, they can be found in great abundance, but where their special needs are not met, it is pointless to look for them. If they could exist within a different environment, they would—but that is not the way they have evolved, and they have long, long ago become completely locked into their requirements. Thus, they are said to be *endemic* to those areas where their vital needs are satisfied. They are found there, and nowhere else.

Hotspots, whether on land or in the oceans, are therefore relatively compact regions with an unusually high rate of endemism. Significantly higher percentages of the living things that inhabit the hotspots are endemic to those areas than is typically the case for most habitats. Whether because the hotspots boast especially plentiful food, or numerous places that can serve as shelters, or are protected from the forces of nature, or contain an array of unique conditions, a remarkable proportion of the living things found there are not found in any other habitat in the world. This situation is the global biodiversity equivalent of putting all your eggs in one basket. It works beautifully, unless you drop the basket. So let us look a little more closely at some of these oceanic baskets of life where so many endemic species are clustered.

HYDROTHERMAL VENTS: LIFE IN THE PRESSURE COOKER

Hydrothermal vents are created where seawater penetrates channels formed by cooling lava flows.[50] The seawater reacts chemically with the lava and then comes back out of the sea floor as superheated water containing compounds such as sulfides, metals, carbon dioxide, and methane.[51] Hydrothermal vents are located throughout the world's oceans, and while most vents are found in areas of sea floor spreading, they also occur in subduction zones and fracture zones.[52] Vents have been found at depths ranging from 980 to 11,800 feet,[53] while most are found at an average depth of about 6,900 feet.[54] Individual vents have a limited life span of perhaps several decades,[55] and it is thought that vent organisms migrate from vent to vent.[56] Of course, all of this information must be considered preliminary and incomplete, inasmuch as these vents were totally unknown to science until 1977. The science of hydrothermal vents is still in its infancy, but the early data are astounding.

While species diversity at hydrothermal vents is relatively low in terms of sheer numbers, with 443 known species at present,[57] these hyperthermophile species are highly unique and endemic.[58] Endemic species comprise 367[59] of these identified species and undoubtedly many more species will be identified as these areas are further explored.[60] These species are unlike anything else on Earth, relying on chemosynthesis, rather than photosynthesis, as their primary means of producing energy.[61] The estimated market value of the commercial utilization of these vent species is potentially at least $3 billion annually.[62] Additionally, some scientists suspect that organisms similar to these were the "cradle of life" from which life on Earth began, and therefore of amazing scientific importance.[63]

It is difficult to overstate the extent to which the vent organisms are unique. The vents are home to bacteria which thrive on hydrogen sulfide (poisonous to most other forms of life).[64] These bacteria live in water so hot (up to 235°F) it is kept from boiling only by the enormous pressures deep in the ocean.[65] Such water temperatures are impossible under ordinary conditions, and yet life thrives in these superheated, highly pressurized, perpetually dark waters. The thermal vents also provide the only habitat for a large tube worm that manages to grow to more than one yard in length without the benefit of either a mouth or a digestive system.[66] Such seemingly unattainable specializations almost certainly hold the key to great advancements in science, medicine, and technology, if and when they are adequately studied.

SEAMOUNTS

There is another unique marine environment where hotspot-like rates of endemism are present. Seamounts, as the name implies, are undersea mountainlike peaks that rise from the ocean floor without breaking the surface.[67] Seamounts (essentially marine mountains) are located throughout the ocean, usually in chains, similar to terrestrial mountain ranges. Most seamounts are located at a considerable distance from any landmass.[68] Seamounts often project upward into zones closer to the surface and function as submerged islands for marine species that would not otherwise be found in the surrounding ocean.[69] In addition to providing a higher and more life-friendly surface in this zone, seamounts deflect currents and create an area of upwelling.[70] This upwelling brings nutrients into the euphotic (photosynthetic) zone, thus producing pockets of food production in areas of otherwise limited productivity.[71] The total number of seamounts is estimated to be in the tens of thousands, but fewer than three hundred have been sampled.[72]

These fragile ecosystems vary greatly in their biodiversity, have a high degree of endemism, and may be centers of speciation where new species evolve. One study estimated that 15 percent of the benthic invertebrates on seamounts are endemic to a particular seamount, but since this study, more than twice as many invertebrates have been discovered in such areas.[73] Seamounts provide a valuable habitat and shelter for immature fish, and they act as aggregation areas for several commercially valuable species.[74] It has also been suggested that seamounts may act as "stepping stones" for transoceanic dispersal of species, as well as vital stopping points for migratory animals.[75] Because seamounts are often associated with heated areas and volcanic activity, hydrothermal vents are also found at some seamount locations.[76]

HOTSPOTS OF LIFE IN THE OCEANS: WHY SHOULD WE CARE?

There is now evidence that oceanic hotspots[77] are the marine equivalent of terrestrial biodiversity hotspots. That is, they are relatively compact pockets of life with a high degree of endemism, even within the vastness of the oceanic environment, which, covering over 70 percent of this planet by area and even more by volume, is by far the largest on Earth.[78] One study found that the marine hotspots we currently know of tend to be located in subtropical waters between 20° and 30° north and south of the equator.[79] It is thought that subtropical waters might be particularly hospitable to marine species because these waters accommodate both cold- and warm-water creatures, with intersecting currents bringing many species together in eddies with layers of different temperatures.[80] These features are often found near prominent topographical structures such as islands, shelf breaks, seamounts, and coral reefs.[81]

We have the most information about coral reefs, and we know that there is a stunning profusion of biodiversity in these badly threatened areas.[82] As we shall soon see, several of the world's coral reefs have already been identified as the marine equivalent of biodiversity hotspots. The biotic richness of coral reefs has been likened to that of tropical forests, and vibrant corals are now known to exist in far deeper waters than was once thought possible, even appearing thousands of feet below the ocean's surface.[83] As mentioned earlier, there is now evidence that there are *not* dramatically fewer species represented in the dysphotic and aphotic zones, although direct observation is difficult to achieve. But even if the old presumptions were true, and numbers of species were drastically reduced as the distance from the surface increased, this most emphatically would not mean that the biodiversity in these deeper zones is any less threatened, or any less important. But why?

The very unusual environmental conditions that prevail in the ocean's abyss, or other deep and remote regions, are such that the species that do exist there must, by necessity, possess some unique evolutionary adaptations. We know that bioluminescence is found among some of the life forms that brave the darkest regions, and they generate their own light where no other light exists. The traits that also enable creatures to withstand unimaginable water pressures and extremes of cold are unlikely to be found among species that inhabit more accessible, more hospitable homes. Especially when one considers that species in the aphotic zone must live in a perpetual combination of total darkness, numbing cold (or, in the vicinity of thermal vents, scalding heat), and bone-crushing water pressure, only a highly specialized combination of adaptations, in concert, could overcome such bizarre stressors.

The adaptations that can defeat these extreme conditions and allow life to prevail could be of great practical value to human beings, both now and in the future. Gene transplantation from these species to others could result in much hardier, less weather-dependent strains of crop plants and other species actively farmed or raised by people. Granted, genetic engineering is a controversial topic,[84] but there does seem to be much promise for responsibly using genetic traits from deep-ocean species to make other species more robust and less reliant on artificial, pollution-generating chemical protections.[85] Advancements along these lines could produce lifesaving new transgenic strains of food sources, as well as species that give us fabrics, building materials, and other key goods. Likewise, and much less controversially, medicines derived from these hardy species might offer dramatic new solutions to previously unsolved health problems for people and their domesticated animals. There could also be entirely new sources of cheap, abundant, nutritious food hidden away in the oceans, of vital importance to our climbing world population.

In addition to these and other examples of the utilitarian value marine biodiversity offers directly to humankind, there are "ecosystem services" of vertiginously towering size. Some vital marine ecosystem services include the regulation of Earth's climate, the sea lion's share of the hydrological cycle, and the breakdown of organic waste products. One estimate places the annual global value of these ocean services at $23 trillion, almost as much as the world's combined gross national product, and approximately two-thirds of all the ecosystem services on the planet.[86] Even if this figure is inflated by two orders of magnitude, it still represents an astonishingly high value well worth our vigorous efforts to preserve.

I would hope that it is unnecessary to spend a lot of time explaining why we should care about a mass extinction unleashing havoc in our oceans today. It seems so obvious. And above all the powerful, utilitarian, "what's-in-it-for-me" arguments, the fact remains that saving these livings things is the *right* thing to do. In this oh-so-sophisticated post-modern age, it is still possible to speak of right and wrong, is it not? As I will soon explain, we human beings are almost entirely responsible *for* the massive die-off in our oceans, so why should we not be responsible *about* it? It is the shop-worn shopkeeper's motto writ large: "You break it, you buy it." All over the world we are "breaking" the vast, ancient, but fragile structure of marine life, and if we do not pay to undo the damage, it will only spread and worsen. We have a potent moral duty to do all we can to halt and reverse the harm we are inflicting upon this most precious and least understood of natural treasures.

But what about the argument that extinction is a natural phenomenon, just one more fact of life, and those species that become extinct in some way deserve their fate, having lost out to better competitors? Some would say, "If these dying species can't stand the heat, let them get out of the lobster pot, because they have it coming—they've lost the game of Life." Now, it is true that extinction is inevitable for all species at some point, just as death is for every individual. The underlying rule, sad but true, is: one life, one death—one species, one extinction. Of course, some species have much greater staying power than others. I, for one, am still awaiting the global demise of the common cockroach, but it seems destined to outlive us all, having already shown off its longevity for hundreds of millions of years. Who would have guessed, had reality TV been available in the Carboniferous Period, that the lowly cockroach would still be vying for the title of Ultimate Survivor some 300 million years later? They were here before the first dinosaurs and have outplayed and outlasted, if not outwitted, them all. But that is not to say that we should be sanguine about our own actions causing and prodigiously accelerating the extinction of thousands of species. That overdrive, human-provoked brand of extinction is anything but natural and inevitable. It is akin to the distinction we draw between death from natural causes and murder.

It has never been a defense to a murder rap that the victim was just going to die at some point anyway. It is the murderer's causation of the death—or our causation of the mass extinction—at any earlier time than would have been the case without abnormal intervention that makes the killing wrong, and someone's responsibility. The same is true of the wisdom of taking necessary medicines and vitamins and leading a healthful lifestyle. It would be a foolish person indeed who would disdain all of these beneficial activities merely because she was born under the universal curse of all living things to die by and by. It matters when and how we die. Likewise, it matters when and how— and how many, and how rapidly—our fellow species suffer their inevitable fate of extinction.

By the same token, it is disingenuous to take refuge in the claim that the disappearing species have been tested and found unfit in the grand global game, "Survival of the Fittest," hosted by Charles Darwin. Although human beings are, like all living creatures, part of nature, our capacity for producing large-scale climate change and habitat destruction is (thankfully) unique to our species alone. Our extreme and far-ranging effects on habitats, both marine and terrestrial, are without parallel in the natural world. The species we are prodding into oblivion were fit enough to survive for millions of years before we turned our modern technologies against them, and their inability to withstand our artificial, warlike assault is in no way a justification to saw off their little branch on the Tree of Life; to suggest that it is would be akin to saying that a murder victim got what was coming to him because he did not wear a bullet-proof vest, had not thought to evolve body armor, or got in the way.

Therefore, there is much merit in safeguarding habitats on the basis of something other than sheer numbers of species alone. Among the most significant ecosystems for preservation may well be those that harbor the fewest, the rarest, and the least "successful" living things. Indeed, some of the competing methods of establishing priorities for conservation are based on the criticism of and alternative to the hotspots paradigm. It may be at least as vital to preserve representatives of many unique, if sparsely inhabited, eco-regions as it is to save those areas with the greatest numbers of endemic species. The "List of Global 200 Ecoregions" is a notable example of this approach because it lists representatives of the most important habitats on Earth in an effort to fairly include and save all major variants in habitat type.[87]

There has long been a badly misguided, but unfortunately widespread, belief that life in the oceans is far less susceptible to extinction than land-based species. The vastness of the oceans along every axis— longitude, latitude, and vertical depth—appears to provide virtually limitless habitat for marine life. And with water available in such staggering quantities, it is easy to see why the traditional bromide came to be, "The solution to pollution is dilution." With that much water, what harm is there in dumping all manner of wastes at open sea in gargantuan amounts? The oceans must be virtually immune to extinctions. With so many places to live and so many niches to fill, how could any species ever run out of suitable habitat? And with all the Earth's oceans interconnected, with no apparent walls, barriers, fences, or other physical obstacles separating them from one another, how could any exploitation of marine food sources constitute over-fishing to a meaningful extent? The oceans are the ultimate in-exhaustible, indestructible resource, the epitome of infinity. Right?

Would that it were so. But, the Academy Award–winning song "Under the Sea" notwithstanding, "life in the muck" is not neces-sarily "in luck." The vastness of the oceans' habitat is illusory, as is the supposed immunity to extinction.[88] For example, many marine species can only survive within a rather narrow range of conditions, dependent on the appropriate light, warmth, water pressure, nutrients, chemical characteristics, physical topography, and proximity to the surface.[89] For species that inhabit one particular area of an ocean, it is of no consequence that there might be many other suitable habitats some distance from their own, separated from their current location by inhospitable territory, because they could not navigate from one area to another. To do so would require the species to endure con-ditions beyond their acceptable limits for survival. Similarly, for the species adapted to a particular depth, with all the attendant parame-ters, it is irrelevant that the water might extend for another 30,000 feet below them; they could never live under those conditions, and all those additional miles of vertical space are utterly off-limits as po-tential habitat. After all, it is no practical consolation to a person squeezed into a crowded tenement that the sky above her has plenty of empty living space for millions of miles into the heavens, or that the dirt beneath her apartment floor stretches down to the very core of the planet. She is still stuck in a 450-square-foot unit with the rest of her family, and that is the reality of her living space.

It is far from true that "the world is your oyster," or that all the world's oceans are a suitable habitat for oysters, or for any other type of marine life. Just as on land, marine species have specific needs as to temperature, food, amount of sunlight, type of terrain or water, and other aspects of their environment. There are certainly idiosyncratic, and even unique, marine environments that provide habitats for species that have adapted to unusual conditions, just as on land. Such ecological niches as coral reefs, kelp forests, seagrass beds, hydrothermal vents, seamounts, and mangrove forests constitute only a very small part of the sea, but are home to huge numbers of life forms specialized for the conditions found only there.[90]

Some marine species are more adaptable to a wide range of conditions than others, and those species tend to be much more ubiquitous than those that are specialized to a very narrow range of conditions.[91] But it is certainly true that many marine species are limited, by their very nature and millions of years of adaptation, to a rather small habitat area, both in terms of latitude and longitude as well as depth of water. When that little area is altered or ruined by human activity, the consequences can be just as devastating as those that follow from adverse modification of terrestrial habitats. We know because we have tried it, and we are seeing the results pouring out on top of us in the form of the sixth mass extinction in our planet's long history.

HOW ARE WE KILLING OUR OCEANS?

There is ample evidence that human activities adversely affect the sea in a variety of ways, some more readily apparent than others. Ocean dumping, introduction of invasive species, development of coastal areas and the attendant discharge of materials into the waters, sedimentation and eutrophication from agriculture and silviculture, and overfishing in a particular area may well have severe impacts on life in the immediate region and often far beyond.[92] Within a given marine locality exhibiting a certain depth, proximity to major currents, ambient temperature, and the like, living things are interdependent and linked in much the same way as are the denizens of any terrestrial ecosystem. When there is a major perturbation of that ecosystem, whether by chemical pollution (organic or inorganic), noise pollution, underwater detonation of explosives, overharvesting,[93] introduction of exotic species, trawling, dredging, sedimentation from runoff, climate change, or any other stressor, a significant decimation of one species will affect others with a nexus to the species in the food web and in the broader array of ecological relationships.[94] In the marine realm, the term "ripple effect" thus has special relevance.

The phenomenon of overfishing and the collateral damage that flows from it is especially pernicious because it has pursued its prey even as the prey retreats.[95] As populations of commercially valuable fish and other sources of seafood have been depleted in one region, fishers have switched their keen attention to previously less-desirable species and individuals and/or have moved their operations to ever more remote and deeper waters.[96] The chase has relentlessly followed the retreating, vanishing biodiversity, moving sequentially from the nearby continental shelf to less accessible, more distant waters and from prized species to more marginal catches.[97] The progressive shift in targeted species is a process known as "fishing down the food web," and smaller, younger, and less-valuable organisms serially take their unenviable turn as the hunted, with profound effects on marine ecosystems.[98] To a significant degree, fishing down the food web has inflicted major changes in the structure of marine food webs and contributed to a global crisis in fisheries, and in marine biodiversity more generally.[99]

This practice of fishing down the food web has also featured shifts to increasingly more technologically sophisticated and/or more deadly means of locating and catching the ever scarcer, ever more distant prey, as I will soon address. Because fishers make these adjustments in their fishing locations, methodology, and targets to compensate for collapses in commercially valuable species in overfished regions, the decline in marine biodiversity can temporarily be masked. Fishing down the web allows the total overall catch to remain constant or even to rise, for a time, as new areas and new targets are exploited with new and more effective techniques. On a superficial level, it can appear that there is no problem because we are catching as much fish (broadly defined) as ever—so where is the mass extinction? But eventually there is no longer any web to fish down, and the aggregate catch declines dramatically, dragging down with it key portions of imploding marine ecosystems.[100]

You may have personally witnessed some evidence of the shift in fish availability within your own lifetime. You need not be quite as ancient as I am to remember a time when the varieties of fish sold in markets and featured on restaurant menus were different from those we see today. Some of the fish once commonly available are no longer plentiful enough to be economically affordable. They have been re-placed, in supermarkets and in restaurants, by other species of fish—fish that not so long ago were disdained as "junk" fish, unworthy of human consumption, or at least insufficiently prestigious to have snob appeal. Yet now these junk fish (such as catfish) are proudly touted as the catch of the day and are offered for sale at fancy prices because we have fished their snootier cousins into oblivion. Reality has intruded on our snobbery. Fame is fleeting, and the types of fish we love to death are continually changing, by necessity.

Because of our abysmal ignorance of biodiversity in the abyss, or in other portions of the dysphotic and aphotic zones, there is a titanic gap in our knowledge of the extent of the extinction threat there. How can we possibly know how many species or even higher taxa have been lost, or are now in danger of extinction, when we have no inkling what was down there to begin with?[101] Is there any way we can accurately gauge the collective impact of human activities on deep-ocean biodiversity, or is this question literally out of our league?

Our technological capability to explore the deep ocean, whether through human-occupied or remotely controlled means, is quite lim-ited. The United States ranks only in fourth place worldwide in terms of our ability to probe beneath the ocean's waters; Japan, Russia, and France are in the lead.[102] The United States' deep-sea submersible, named *Alvin*, can operate at a maximum depth of 14,764 feet.[103] But even the most advanced marine exploration craft in the world is not capable of getting anywhere near the benthic area of the deepest waters. Until the capability of exploring tens of thousands of feet deeper than ever before is developed, no one can do more than guess the state of deep ocean biodiversity. It might be said that all we know is what we can see through a glass, darkly, but we do not even come close to that level of access.

This is an appalling lack of information on perhaps the most vital issue facing the Earth today. We are left with little more than anec-dotal evidence from those small segments of marine life visible to us, which we can then attempt to extrapolate into the unconquered, everlasting midnight of the unexplored depths. This is wholly un-satisfactory, from both a scientific and commonsensical standpoint, but it is all we have. This is the legacy we have reaped from our collective failure to make ocean exploration a priority on anything remotely approaching the scale of our space program. I do not begrudge anyone the money and effort we have expended to travel to and learn about the moon and the rest of our solar system. That is important and fasci-nating work, and we should continue investing in it. But we are living through, and dying from, the payback attached to paying so little at-tention to the habitats that make up most of our own home planet. In our passion to break free from the gravity that anchors us to our Earth, we have been too quick to turn our backs on the oceans before we ever even truly faced them and saw what they hold.

What evidence can we glean from this embarrassment of poverty? The signs clearly point to a mass extinction in the world's oceans, something long thought impossible.[104] According to a World Con-servation Union (IUCN) study, fish are now the most vulnerable of all groups of living things, with up to one-third of all known fish species threatened with extinction.[105] Some noteworthy and startling studies suggest that the global oceans, in the aggregate, have lost more than 90 percent of all large predatory fishes during the industrial period when highly destructive longline fishing methods became widely used.[106] Industrialized fishing has been blamed for 80-percent re-ductions in marine community biomass in many cases within fifteen years of the onset of intensive exploitation.[107] Of course, the whole-sale removal of top predators from any ecosystem can be expected to have far-reaching effects beyond the predators themselves. Their dis-appearance causes a cascading or domino effect, leading to a sim-plified, impoverished, and much more vulnerable ecosystem at many levels.[108]

We are more intimately familiar with the results of people ex-terminating the dominant carnivores in terrestrial habitats, and our experience can assist us in gaining an understanding of what is hap-pening in the oceans. Our slaughter of wolves, mountain lions, and other large predators has knocked many ecosystems into instability, with meteoric population changes among the animals that were long kept in check by these natural hunters. Previously stable herbivore populations have soared to the point where creatures like deer and moose are now serious health and safety threats to humans. Those same populations have sometimes crashed back down because of drastic food shortages or rampant disease. Similar secondary and tertiary effects follow our remorseless overexploitation of large, predatory ocean fish. When we drive these marine carnivores within 90 percent of the extinction line and beyond, we force major and dire changes in the ecosystems they once held together, and no one knows exactly where or how those changes will eventually sort themselves out.

Climate change has also been linked to serious loss of marine biodiversity, notably in the coral reefs and very likely in many other ocean habitats as well. According to recent research, water temperature in the tropical oceans has increased by nearly 1°C during the past hundred years, and is currently rising at a rate of about 1°C–2°C per century.[109] This climate change has caused, in conjunction with other human-made stressors such as overfishing, mining, and sedimentation, between 50 and 70 percent of all coral reefs to become seriously threatened. There have been at least six major episodes of coral "bleaching" since 1979, with massive mortalities in the species found in the reefs. Entire coral reef systems have died following such bleaching events.[110] Given their extraordinary significance as marine centers of endemism, this extreme degree of coral reef ruination can only be judged catastrophic. Of all the ocean habitats, coral reefs are probably the ones we can least afford to lose, yet that is precisely what is happening, on a dreadful scale.

Scientific research indicates that, despite (or more likely because of) large increases in total global fish catches in recent decades, overfishing is a major problem.[111] There is evidence that about 10 percent of the world's major fisheries are depleted, 15–18 percent are overexploited, and 47–50 percent are fully exploited. This leaves only about 25 percent of the world's marine fish populations at the relatively healthy underexploited or moderately exploited level of intrusion.[112] Indeed, some important fisheries have collapsed altogether, resulting in emergency closures.[113] And as certain waters become depleted of fish, commercial fishing operations move to other regions, spreading the devastation into ever more remote parts of the ocean.

The net result of all of this aggressive harvesting is shocking. At least 90 percent of each of the world's large ocean species, such as cod, halibut, tuna, swordfish, and marlin, have vanished during recent decades.[114] Since the advent of large-scale modern fishing in the 1950s, it has taken an average of only fifteen years to reduce by 80 percent or more any fish species targeted.[115] For some species and some regions, the population crash can be even more precipitous.[116]

Part of the reason for the carnage is technological advancement in fishing techniques. With tracking buoys, lightweight and nearly invisible nylon nets, satellite data, and sophisticated sonar, it is now possible to locate and catch, efficiently and regularly, previously elusive concentrations of fish. We have learned to use devices and techniques such as dynamite fishing, muro-ami,[117] poisoning, otter trawls, beam trawls, scallop dredges, clam dredges, and St. Andrews' crosses, to disastrous effect.[118] It is not surprising that systematic use of toxins, explosives, and bottom-scraping methods have a less than salutary effect on marine biodiversity. Suffice it to say that when we decide to hunt or fish for commercial profit, we are very good at killing and breaking. To paraphrase boxing great Joe Louis, "The fish can swim, but they can't hide!"

The high-tech, satellite-and-sonar, stealth mode of commercial fishing today has strong parallels with computer-age warfare. It is as if we have World War III on our hands, and this time all the fishing nations of the globe have become allies and have in unison declared war on marine life. During our recent human-on-human wars, including the fighting in Iraq and Afghanistan, we have seen the deadly use of sophisticated tracking and guidance systems, supported by superb satellite photographs, computer-aided targeting, unmanned intelligence-gathering aircraft, Global Positioning System precision location methodology, and advanced radar and sonar technology. This splendid information collection and dissemination apparatus enabled the United States and its allies to direct the most advanced, precise, and powerful arsenal of smart bombs, Stealth bombers and fighters, cave-buster explosives, and finely guided missiles the world has ever witnessed. The conventional forces arrayed in opposition to this latter-day juggernaut never had a chance.

In World War III, the War on the Water World (W3), we are armed with very much the same detection, tracking, hunting, and killing weapons we have used to find, pursue, and destroy our two-legged enemies in the Middle East. Although the objective in W3 is not only to find and kill, but also to bring back the dead bodies of many selected victims for our dining pleasure, the same Silicon Valley technologies are being deployed, and the targets are being located and "taken out" with success rates that were unheard of until very recent years. The age-old catch-as-catch-can fishing methods that often came up empty after weeks of lonely searching at sea and bone-wearying hard work have suddenly been supplanted by every orbiting eye-in-the-sky and computer-chip enhancement we have used to such phenomenal effect to root out and exterminate our military foes. It is no surprise, then, that submerged hiding places and de facto sanctuaries that were reliable for thousands of centuries have been turned, in just the last few years, into target-rich fishing ponds ready for the taking. The harvest, or slaughter, of marine life now abruptly exposed to the full fury of Star Wars–caliber military prowess is at a level never before dreamed of by the most visionary fishermen, nor faced in the most horrific nightmares of conservationists.

Of course, a lot more living things are caught, and killed, as part of our space-age commercial fishing operations than just the targeted species.[119] When fishers use trawl nets large enough to snare twelve Boeing 747 jet airliners simultaneously, one should not expect a high degree of selectivity in the catch.[120] About 27 million tons of "by-catch" or "by-kill" creatures are destroyed every year as collateral casualties in the hunt for the most desirable fish. If this indiscriminate slaughter is not an immense case of friendly fire, it is at least indifferent and apathetic fire on a world-record level. It is estimated that at least one ton of living things is killed as by-catch for every three tons of marketable fish caught.[121] Where the practice of "high grading" is employed, discarding all except the biggest and most valuable fish so as to derive maximum dollar value from the limited hold space on fishing vessels, even larger percentages of the total kill are needless, valueless by-catch.[122] This unfathomable waste is associated with many modes of industrialized commercial fishing, but it is especially pronounced when one late-twentieth-century "improvement" in fishing methodology is used.

Probably the most blindly devastating weapon in the modern arsenal of high-tech/high-wreck commercial fishing has been the drift net, and its close relative, the longline (which kills with hooks as well as entanglement). Drift nets are enormous nets up to forty miles long and between twenty-five and fifty feet from top to bottom. Often made of transparent, nearly invisible nylon mesh,[123] drift nets hang suspended in the water, with floats on the surface and weighted lead lines. This arrangement allows the immense nets to hang straight down in the water for extended periods during which they are left unattended to do their dismal work passively, without any human intervention, with awful impact on "non-target" living things.[124] Fish and other marine life trapped in the nets die in huge numbers from starvation, strangulation, asphyxiation, and a variety of related and often wretchedly cruel causes. In fact, there has been a secondary problem as damaged portions of abandoned drift nets continue to haunt the oceans as "ghost nets" long after they have ceased to be monitored by people. But never have any ghosts done as much damage, or been as terrifying, as these all-too-solid nets. When we talk about the "net result" of something, we generally do not think in the literal sense of the carnage wreaked by drift nets, but there is no net result anywhere more appalling.

These "curtains of death" exact a dreadful toll, not only on fish but also on birds,[125] sea turtles, and marine mammals such as dolphins, whales, and sea lions, which often feed on species targeted by the drift nets and are accidentally trapped in them as by-catch. The unintentional decimation of such important nonfish marine creatures adds another layer of rubble to the tumbling structure of ocean ecosystems. When the drift nets are anchored on the ocean floor, they also inflict great harm on deep sea corals, sponges, crinoids, and other habitat-creating life forms. And because drift nets and longlines are frequently (and very deliberately) sited along major fish migration routes, in the biodiversity-rich upper levels of the euphotic zone, they are instruments of killing as effective as they are indiscriminate. Due to public outcry leading to United Nations resolutions, the use of the largest drift nets in the open ocean has now been somewhat reduced, but by no means totally eliminated, as recent studies make all too clear.[126] Further, there remains a serious threat from smaller drift nets and longlines used in coastal waters, and terrible losses are still being inflicted. It is amazing and disheartening how proficient we are at killing things we are not even aiming at. In our oceans today, we thus have the tragedy of, not a drive-by shooting, but swim-by trapping, on a scandalously massive scale, by hook or by net.

Some other modern (and semimodern) commercial fishing methods cause the direct destruction of key marine habitats, in addition to taking large amounts of life as intentional catch and collateral bycatch. Trawls, dredges, poisons, and dynamite (yes, dynamite) severely damage the seabed environment, eliminating hiding places, living spaces, and other refuges for marine life forms, and eviscerating deep-sea corals.[127] Some deep-sea cold-water corals that have been gouged by trawls are estimated to be 4,500 years old, yet they can be destroyed in a single night of trawling.[128] This appalling and wanton waste is almost impossible to describe in words. It is roughly equivalent, in an ecological sense, to such acts of cultural barbarism as the Taliban's despoiling of several gigantic and ancient statues of Buddha a few years ago. Indeed, few phenomena are as sickening to civilized people as the ability and eagerness of some humans to ruin in an instant that which took so long to build. Natural wonders older than the pyramids are being wiped out with abandon this very day as we abandon our natural heritage.

The habitat modification is large-scale and horrifically harmful, especially to fragile and highly complex benthic niches that cannot withstand this type of brutal physical disturbance.[129] Of course, not all marine habitats and marine species are equally disrupted by demersal trawling, and some types of trawling gear are more harmful than others. In benthic habitats, regions such as the areas around seamounts and previously unexploited depths are more vulnerable than simpler seabeds on the near-shore continental shelf. Likewise, the various types of shellfish dredges, rock-hopper otter trawls, and heavy flatfish beam trawls cause the most intense disturbance to the seabed, while lighter gears such as smaller otter and prawn trawls are not usually as destructive.[130]

This is something we are really sinking our claws into. Dredges and heavy trawls inflict especially drastic habitat modifications. They are dragged along the seabed, typically converting ecologically rich, complex habitats to much simpler, biologically impoverished areas as they alter the benthic surface topography, churn up and resuspend sediment, and induce changes in biogeochemical processes.[131] Coral reefs, seamounts, hydrothermal vents, and many other ocean-bottom communities that might have required centuries or millennia to develop are scraped, raked, broken, shattered, leveled, buried, and obliterated by these trawls we deploy with blind and barbaric thoughtlessness. The effect is akin to plowing under a tropical rainforest.

Additionally, these and other modern fishing techniques kill large numbers of marine species that are never even trapped in a net or hooked on a line, thereby indirectly excising vital portions of the marine food web through habitat destruction.[132] And, as I indicated earlier, we now can, and do, drag, poison, blast, and otherwise disrupt important marine habitats even in remote regions that were previously beyond the limits of human technological capability, turning what had once been de facto marine wildlife refuges into eminently ex-ploitable "shooting fish in a barrel" fisheries. Put another way, our technological progress has enabled us to expand our operations into portions of the ocean that had previously been out of reach and im-possible for humans to target effectively. The ultimate kinds of marine protected areas—the areas we simply were incapable of exploiting, and could not get to no matter how much we wanted to—are now targets of opportunity right in our crosshairs.

Overfishing thus has several important and interrelated influences on marine ecosystems and their biodiversity. To summarize, these influences include the direct removal of target species; direct changes in size, age, and fecundity structure of target populations; alteration in and reduction of nontarget populations; changes in the physical en-vironment; modifications of the chemical environment, including nutrient availability; and trophic cascades.[133] If we want a mass ex-tinction, this is a fine strategic plan.

A major factor contributing to the overfishing debacle is that governments all around the world have decided to become living exemplars of Malcolm X's famous admonition: They are part of the problem, not part of the solution. By subsidizing their commercial fishing industries with up to $20 billion a year worldwide, many governments ensure that even unprofitable, needlessly destructive, indiscriminate, and unsustainable fishing operations continue.[134] Governments that prop up shaky fishing industries to benefit their own local economies do so at the expense of Earth's marine biodiversity— the tragedy of the commons towed out to sea.[135] Indeed, not only in overfishing, but also in ocean dumping, oil/mineral/gas exploration, and other practices, the international waters of the open ocean represent an archetypal case of the tragedy of the commons. A "global" resource, part of the common heritage of all humankind, owned by everyone and no one, has been exploited, abused, and neglected to a shameful extent, because each nation sees no reason to exercise restraint in manifesting its maritime destiny while so many others are throwing caution to the waves.[136]

This accumulation of evidence shows that, as with terrestrial hotspots, vital pockets of endemism in the oceans are threatened by habitat destruction or alteration of various types. In addition to the domino effects (trophic cascades) on food webs spurred by overfishing and needlessly indiscriminate fishing methods,[137] habitat decimation and modification also stems from ocean dumping and pollution, indirect contamination, sedimentation, and eutrophication brought on by land-based activities, the effects of oil or mineral exploration and exploitation, and changes in temperature and salinity caused by climate change.[138]

Over 80 percent of ocean pollution results from land-based activities, which may be surprising to people who are accustomed to thinking of all marine pollution within the stereotype of the leaking oil tanker.[139] Every eight months, over 11 million gallons of petroleum enter the marine environment from dry land in the form of runoff. This is equal to the entire amount of oil spilled by the *Exxon Valdez* tanker during its infamous incident. But, just as mass extinction happens far more slowly and invisibly than we might expect, the main source of marine pollution is not huge, headline-grabbing disasters at sea, but rather a daily diet of deadly stuff trickling and drifting down from the land and the air in countless places. Contaminants have been found in the middle of the ocean far from any shore and in deep ocean-floor sediments.[140] Nitrogen, other nutrients, and toxins from runoff and atmospheric sources (such as agricultural fertilizers, animal wastes, municipal sewage, smokestacks, and automobile exhaust) enter the ocean waters and can cause eutrophication, an overfertilization of the marine environment.

In extreme cases, eutrophication and the huge toxic algal blooms (including "red tides") it often generates can contaminate seafood, upset the balance of life forms in the oceans, poison people,[141] and spark the literal conversion of ocean into dry land. The overfertilization depletes oxygen in the waters as marine plant life (mostly algae and phytoplankton) proliferates. This can lead to an anoxic environment, such as the 7,000-square-mile (and growing) "dead zone" area at the mouth of the Mississippi River.[142] An excess of nutrients can also cloud the waters, lessening the sunlight that penetrates the euphotic zone, and the many life forms that depend on a high level of sunlight may experience disastrous die-offs.[143]

Land-based activities also contribute indirectly but formidably to the harmful modification of marine habitats through means other than "conventional" pollution. When terrestrial forests are cleared, there is often a large increase in development and in soil erosion. With the forests gone, pollutants and sediments flow into the coastal waters in higher quantities, leading to many of the impacts on marine habitat I have described.[144] This is one of the most significant and insidious of the myriad ill effects associated with deforestation. It is one more example of how interconnected Earth's habitats are, and how far the deleterious consequences of our destructive actions can reach. When we deliberately eliminate one habitat on land, we may have no idea that we are simultaneously contributing to the ruin of a very different but equally vital habitat in the ocean—that is what can happen in the highly interrelated web of life. It is especially appalling that the devastation of one type of hotspot (the tropical forest) can lead to the ruination of another type of hotspot (the near-shore coral reef) as we kill two hotspots with one stroke. An old cliché states that no man is an island, but the reality is that no island is just an island either.

The remaining 20 percent of ocean pollution is a result of human-caused marine activities, including collisions, accidental discharges, and deliberate operational discharges. These releases can have severe localized impacts. For example, the Galicia Bank is a large seamount encompassing about 2,400 square miles in the North Atlantic. In November 2002 the damaged tanker *Prestige* sank in the vicinity of the

Galicia Bank, spilling some 50,000 tons of heavy fuel oil at the slope of the seamount.[145] For such a priceless ecological treasure, an incident of this type is phenomenally destructive, but the sad truth is that the seamount was in the wrong place at the wrong time.

Advances in oil and gas drilling technology place many of key marine areas at risk. In the early 1960s the water depth limit for exploratory and production drilling was 980 feet and 328 feet, respectively.[146] By the early 1990s these limits increased to 6,500 feet for exploration and almost 3,300 feet for production.[147] The current exploratory drilling record is 8,015 feet and is expected to rise to 9,800 feet in the near future.[148] These increases are being driven by the tremendous amount of oil that is believed to exist under the deep ocean floor, an attraction with a siren song that people have found irresistible.[149]

Additionally, possible mining of methane hydrates, a potentially lucrative energy source, could severely impact the vent and seamount communities. Methane hydrates, a combination of methane and water, have been found on the seabed and embedded up to 1.8 miles *under* the seabed.[150] The energy equivalent of the worldwide deposits of methane hydrates is estimated at twice the amount of all other fossil fuels on the planet![151] It does not take an economic genius to deduce that the prospect of such a Scrooge McDuck money-bin-filling payday is enough to incentivize veritable swarms of opportunists to do whatever it takes to grab all that wealth. Inevitably, any exploration and extraction activities in pursuit of oceans of energy industry profits will have an adverse impact on these irreplaceable marine communities. As the prospects for immense riches improve, our intrusion will grow increasingly frenzied, and the despoliation of the affected areas will kick into full throttle.

Some seamounts and hydrothermal vents also contain valuable mineral deposits which could make them potential mining sites.[152] Hydrothermal vents can feature deposits of copper, manganese, and gold.[153] While many plans for ocean mining remain just plans, there is at least one operation which may soon begin production.[154] In 1997 the government of Papua New Guinea granted Nautilus Minerals Corporation a license to explore for minerals in the Bismarck Sea, an area totaling more than 1,900 square miles.[155] One only has to look at the damage that humans have done to the terrestrial environment in pursuit of minerals to picture the eventual results of mining in marine areas, as soon as such mining becomes economically feasible.[156] I wonder what the marine equivalent of strip-mining will be. Whatever its precise form, I think it is safe to predict that it will do great harm to the fragile and unique habitats on which it will be inflicted.

Although the issue of environmental impacts from deep-seabed mining requires much more scientific investigation before the full magnitude of the problem is known, there are undoubtedly several ways in which such mining, as well as drilling, and oil/gas exploration, can have an adverse effect on marine biodiversity. The most significant impacts will probably be caused by the physical disturbances from the act of mining, drilling, and collecting, in which soft sediment will be disrupted, shifted, or compressed while hard substrates (which are home to many organisms) are torn away and deep-sea corals are obliterated.[157] There will also be discharge of wastewater and noise from the mining ship and its equipment.[158] And if the processing of mined materials takes place near the shore, side effects from these activities could return to the water in the form of harmful solid, liquid, and gas emissions.[159]

There is a similar risk related to a very different kind of marine treasure-seeking: the hunt for rare and valuable combinations of DNA. Given the unique species that inhabit these areas and the enormous profit potential of their genetic resources, they are likely to be subjected to unsustainable and haphazard harvesting and exploitation.[160] Removing species from these environments, without fully understanding the functioning of the ecosystems they harbor, could disrupt essential ecosystem synergies resulting in irreversible loss of species. Also, the very act of finding, chasing, and collecting selected living things in these remote areas is bound to be accompanied by collateral damage to the surrounding habitats and neighboring species. Our clumsiness, imprecision, and sheer physical intrusion will cause plenty of harm to species and living spaces other than just the specific creatures being sought.

Our War on the Water World wears many types of camouflage and often lays waste to unintended casualties. We have a wide repertoire of practices that shift the relationships of species within food webs, often eliminating top predators and thereby severely altering the structure and dynamics of an ecosystem.[161] For example, California fishermen attempted to eliminate the California sea otter, which they viewed as competition in the hunt for fish.[162] This caused a rise in the population of sea urchins, which had composed a part of the sea otter's diet. The increased sea urchin population then caused a decline in the kelp, on which the urchins once fed. This decline in kelp in turn led to a decline in the fish population because the kelp had provided critical breeding habitat for many fish species. In retrospect, it is now clear that the sea otters had acted as a keystone species, assisting in balancing the various populations within the ecosystem and, once removed, the entire living edifice crumbled. But that is the way war is. We frequently reap a bitter harvest in the wake of initial "success" in the killing fields.

Unique and fragile ecosystems, such as seamounts and hydro-thermal vents, could be greatly damaged by the overexploitation of certain keystone species. For example, orange roughy is a species of fish that dominates some seamounts in the Pacific Ocean. These fish are extremely long-lived and have a very slow reproductive rate.[163] They have been overexploited as a food resource, resulting in a severe decline in their numbers.[164] In light of how little we know about these areas and what we have learned about keystone species, the effect of this overexploitation could amount to devastation of individual sea-mount ecosystems.

Given that seamounts are generally isolated from each other, it is likely that seamount habitats and animal populations recover from disturbances only over long time periods, if at all, through sporadic re-colonization from nearby seamounts and continental shelf areas.[165] Where this recolonization is disrupted, excessive removal of seamount species may lead to their local extinction and perhaps their global ex-tinction.[166] And when we act in ignorance, as we so often do, we can easily tear a keystone species out of its ecosystem without even knowing it, causing an ecological collapse far out of proportion to the immediate loss of just one species, and irreparable harm to the marine community.

The International Council for the Exploration of the Sea (ICES)[167] has cautioned against the overfishing of stocks by introducing dras-tic catch reductions and placing a temporary ban on the use of bottom-trawling gear in sensitive deep-sea areas. In 2001 the Conven-tion for the Protection of the Marine Environment of the North East

Atlantic (OSPAR) placed seamounts on its priority list of threatened species and habitats of regional concern in order to develop measures for their conservation.[168] Although scientists from ICES had recommended that OSPAR withdraw seamounts from this list due to lack of specific scientific data on actual threats,[169] as of this writing, OSPAR has kept all seamounts, wherever they occur, on its priority list.[170]

Finally, I want to mention the threat of climate change. Changes in atmospheric and ocean circulation attributable to climate change could have very significant adverse local, regional, and global implications for the distribution and abundance of living resources in the oceans. Climate change can interact with all the other stressors I have mentioned, even in unpredictable ways, synergistically or additively. The ocean has a system of currents that redistribute heat, chemicals, and nutrients throughout. Increases in temperature may slow or shut down the Gulf Stream, causing vast changes in species distribution and increasing the risk of hypoxia in the deep ocean. Although subtle, slow-acting, and difficult to assess, climate change (such as global warming) could ultimately prove to be the most important factor in our current Sixth Extinction. Unfortunately, we probably will not know for certain exactly what the dimensions of the problem are until it is quite irrevocably too late to do much about it.

There are other threats to marine biodiversity as well, such as increases in ambient noise. I will not go into detail here, because the other dangers are so profound and tend to dwarf the secondary harms, but even the addition of high levels of noise to an ocean area previously peaceful and calm can have a major impact on the local life. One distressing noise-producing threat is posed, ironically, by visits to these areas by ecotourists. One report indicates that an environmental company is offering tourist trips to visit hydrothermal vents off the Azores.[171]

WHERE ARE THE MARINE HOTSPOTS?

What, then, is the net effect of all of these environmental assaults on marine biodiversity? As mass extinction proceeds, species have vanished from their former habitats, and often stage a "last stand" in certain limited niches, clinging to survival. These last-stand areas, or hotspots, are the key to arresting the extinction spasm. It is imperative that, at a minimum, we prevent further ruination of these vital centers of endemism to forestall even more extreme acceleration of the current extinction rate. Hotspots are loaded with species we cannot find anywhere else in the world. They are *all we have left* of utterly huge portions of life on Earth. We simply cannot afford to lose these delicate treasure chests of biodiversity. But where are they?

Prominent nongovernmental organizations (NGOs) such as Conservation International (CI) have championed certain high-priority marine areas that deserve particular attention. The focal point for CI's recent work has been coral reef hotspots because, as alluded to earlier, more is known about these near-shore areas and they are undeniably among the most important and endangered zones of life.[172] Also, these coral reefs are relatively small, and are home to large numbers of endemic species, making them acutely vulnerable to extinction pressures and a natural focal point for conservation priorities. Indeed, there is evidence that 25 percent of the world's coral reefs have already been destroyed or severely degraded by stressors arising from global warming alone, in addition to immense damage perpetrated by intensive fishing and sedimentation from runoff.[173] The following are the top ten coral reef hotspots as listed by CI, ranked according to degree of threat.[174]

1. Philippines
2. Gulf of Guinea
3. Sunda Islands
4. Southern Mascarene Islands
5. Eastern South Africa
6. Northern Indian Ocean
7. Southern Japan, Taiwan, and Southern China
8. Cape Verde Islands
9. Western Caribbean
10. Red Sea and Gulf of Aden

Eight out of these ten coral reef hotspots are adjacent to terrestrial hotspots, as recognized by CI. In the aggregate, these ten marine hotspots add up to only 0.017 percent of the oceans' area, yet they contain fully 34 percent of all restricted-range coral reef species—a remarkably disproportionate concentration of imperiled species.[175] This is a classic illustration of the hotspots concept: relatively limited area, but tremendous concentration of endemic and endangered species. Coral reefs such as these are vulnerable to on-land human activities such as agriculture, deforestation, and other development that causes large quantities of sediment, nutrients, and chemical pollutants to enter the coastal waters.[176] Of course, overfishing, dredging, trawling, and climate change are also major threats to coral reef hotspots.

Another NGO, the World Wildlife Fund (WWF), has included an extensive representative array of vital marine habitats in its Global 200 list of key eco-regions.[177] The WWF groups their marine eco-regions into five main categories of habitat: Polar, Temperate Shelf and Seas, Temperate Upwelling, Tropical Upwelling, and Tropical Coral.[178] The marine eco-regions are as follows.

Polar:

1. Antarctic Peninsula and Weddell Sea
2. Barents-Kara Seas
3. Bering Sea

Temperate Shelf and Seas:

1. Chesapeake Bay
2. Grand Banks
3. Mediterranean Sea
4. New Zealand Marine
5. Northeast Atlantic Shelf Marine
6. Okhotsk Sea
7. Patagonian Southwest Atlantic
8. Southern Australian Marine
9. Yellow Sea

Temperate Upwelling:

1. Agulhas Current
2. Benguela Current
3. Californian Current
4. Humboldt Current

Tropical Upwelling:

1. Canary Current
2. Galápagos Marine
3. Gulf of California
4. Panama Bight
5. Western Australian Marine

Tropical Coral:

1. Andaman Sea
2. Arabian Sea
3. Banda-Flores Seas
4. Bismarck-Solomon Seas
5. East African Marine
6. Fiji Barrier Reef
7. Great Barrier Reef
8. Greater Antillean Marine
9. Hawaiian Marine
10. Lord Howe-Norfolk Islands Marine
11. Maldives, Chagos, Lakshadweep Atolls
12. Mesoamerican Reef
13. Nansei Shoto
14. New Caledonia Barrier Reef
15. Northeast Brazil Shelf Marine
16. Palau Marine
17. Rapa Nui
18. Red Sea
19. Sulu-Sulawesi Seas
20. Southern Caribbean Sea
21. Tahitian Marine
22. West Madagascar Marine

The list nicely symbolizes the special status occupied by tropical coral reefs. A glance at the list immediately reveals that there are many more eco-regions in the tropical coral reef category than in any other marine habitat. In fact, there are as many tropical coral reef eco-regions identified as all other marine eco-regions combined.

The WWF has chosen to focus on these key locations because of their uniqueness and their particular endemic flora and fauna, not necessarily because of there is a greater degree of threat to them. However, many of the marine eco-regions undoubtedly are at significant risk due to their proximity to heavily populated land masses.[179] This, in conjunction with the features that set these habitats apart from all others, makes them worthy of special and urgent attention in any legal proposal to preserve marine biodiversity. In fact, by not insisting on a high degree of threat as a prerequisite to inclusion on its high-priority list, WWF has shown that it is possible to be proactive in our conservation efforts and not wait until the situation is desperate. This could be very helpful in focusing attention on vital marine habitats prior to the crisis stage, before too much damage is done to make it possible to restore them fully.

Given that there are such vital epicenters of marine biodiversity, the best way to preserve them is to treat them like the watery equivalent of terrestrial protected areas (such as wilderness preserves, wildlife refuges, and national parks). Traditionally, ocean-based preserves have been called marine protected areas (MPAs).[180] The basic concept is simple, although the implementation can be quite complex: Identify portions of the ocean containing unusually important concentrations of biodiversity, and then establish these as MPAs, with more or less strict controls on allowable human activities in and near them. As we shall see in the following chapters, this promising idea has, thus far, achieved far less than it otherwise might.

It may be somewhat belated, but the United States has begun to focus some official attention on the critical issues outlined in this chapter, and that is better than pretending that all is well. As called for in the Oceans Act of 2000,[181] the U.S. Commission on Ocean Policy released its "Final Report" in 2004 on the condition of the oceans and what the United States should do about it.[182] This massive document could be the beginning of some positive changes at the highest levels of government, but only if the right people read it with the right frame of mind. In that regard, I hope that my book will help provide a useful context for the report, and contribute to a sense of urgency in our efforts to find an appropriate blueprint for the twenty-first century.

With this background in place, let us now examine the ways in which the world's embattled oceans have been categorized and regulated by the law. The legal division of Earth's waters is quite different from the naturalistic zones we have previously considered.

TWO

Law of the Sea and in the Sea

There are several major international treaties and conventions that address in a significant way the issue of marine biodiversity. Most people would expect international law to be the primary means of safeguarding marine biodiversity, particularly in the deep ocean waters far from land, so it is useful to consider the efficacy of international legal protections at the outset. There is an obvious connection: international territory equals international law. The reality is less obvious, but it is worthwhile to explore how our assumptions can be a mirage leading us far from where we really need to go. From here we will move on, in chapter three, to the internal domestic laws of several nations with a special claim to some key marine hotspots.

I should emphasize that this is *not* an exhaustive treatment of the international laws that touch on marine biodiversity. I have chosen to cover in some detail only a few of the most significant international legal instruments, but entire books have been devoted to several of these, and there are others that are also of some relevance that I will not discuss.[1] There are also several regional multinational agreements, in addition to those that are global in scope, that address the issue of marine biodiversity.[2] However, the main points I will make regarding the effectiveness—or ineffectiveness—of the agreements covered in this chapter are also generally applicable to the other global and regional agreements.

I hate to spoil the surprise by revealing the ending in the beginning, but I will do it anyway. I will give you my opinion, which amounts to heresy in the halls of legal academia. I think that the sum total of all the phenomenally voluminous international agreements related to marine biodiversity is about as meaningful an answer to our mass extinction crisis as professional wrestling. There is a lot of pretense, a lot of loud and angry words, and a lot of nothing. It will require the remainder of this long chapter to show you what I mean. I apologize for the dryness of some of this material, tied as it is to the texts and contexts of lengthy international laws, but I have to plunge into the dry stuff for a while to prove my case that this approach is all wet.

UNITED NATIONS CONVENTION ON THE LAW OF THE SEA OF 1982

The Third United Nations Convention on the Law of the Sea of 1982 (UNCLOS, or Law of the Sea Treaty)[3] is the only globally applicable international law dealing with *all* facets of the marine realm, including biodiversity. It has many shortcomings, but it is by far the most comprehensive and potentially influential treaty in existence in the marine setting. It is also certainly one of the lengthiest legal agreements in history, consisting of 320 articles and 9 annexes. If size does matter, and if length alone were any guarantee of legal potency, our troubles in the oceans would be gone forever. Sadly, it takes more than simply spilling a lot of ink on a lot of pages to loosen the knot of problems we have tied ourselves into.

We must be clear about one crucial point from the outset: By no means is UNCLOS exclusively, or even predominantly, an environmental or biodiversity treaty. Its vast, sprawling reach touches on many other areas, from freedom of navigation to piracy, with heavy emphasis on commercial interests in the world's oceans. These multitudinous portions of the treaty are well beyond the scope of this book. And, although UNCLOS devotes one entire portion (Part XII)[4] to the protection and preservation of the marine environment, it also contains numerous other biodiversity-relevant provisions scattered throughout its prodigious length. The rather bizarre organizational structure of the convention thus presents a challenge to anyone attempting a coherent analysis, but I will try to summarize its key biodiversity-related points in as logical a manner as possible.

Overall, the Law of the Sea Treaty/UNCLOS is aimed at shifting certain long-standing positions regarding the oceans on a wide variety of subjects (such as the old notion that all nations should be able to "enjoy" the freedom to pollute the seas). Under UNCLOS, nations have the very different, if ambiguous, obligation to exercise diligent control over marine pollution from all sources. However, the myriad compromises embodied in the convention are nicely represented by one simple juxtaposition. The convention sets down, in Article 192, the "general obligation" to protect and preserve the marine environment, while recognizing in the very next provision (Article 193) the "sovereign right of States to exploit their natural resources," which right must be exercised "in accordance with their duty to protect and preserve the marine environment." Can a sovereign right to exploit natural resources coexist with a general obligation to protect and preserve the environment? The two seem inconsistent, if not blatantly contradictory. But, as Al Jolson used to say, 'Settle back, folks, 'cause you ain't seen nothin' yet!"

UNCLOS is one of the most controversial of all international laws, in addition to being the most ambitious and comprehensive of the laws governing the world's oceans.[5] Unlike many other international agreements, UNCLOS does not allow parties to make formal reservations with regard to certain provisions; it is an all-or-nothing, take-it-or-leave-it proposition. The United States has chosen to vote with its flippers and "leave it," and has never come on board, despite helping itself to useful territorial/jurisdictional provisions.

The history of UNCLOS is littered with the flotsam and jetsam of many years of contentious, highly politicized debate.[6] Issues of national self-interest, such as the right to engage in and profit from deep-sea mining, led some economically advanced nations such as the United States to remain on the sidelines, frightened off by the prospect of all nations sharing in revenues derived from deep seabeds outside national jurisdiction.[7] It was not until November 16, 1994, after twelve years of bitter struggles, that the convention finally garnered enough signatories to be enforced.[8] Nevertheless, UNCLOS has ultimately attracted numerous signatories, with a total of 148 as of August 2005.[9] Moreover, even nations (including the United States) that have refused to sign on have taken some actions along lines specified by the convention, such as declaring an expanded territorial sea or an exclusive economic zone.

UNCLOS numbers among its multifarious contributions a widely accepted scheme for establishing jurisdiction in the marine setting. The major legal subdivisions of ocean waters are predicated on proximity to a sovereign nation. Under UNCLOS, and most likely also under customary international law,[10] an individual nation's right (or jurisdiction) to control the resources in a given marine area depends, in part, on the physical characteristics of the ocean floor.

Under UNCLOS, any coastal nation has three distinct means of exerting jurisdiction over an ocean area. First, under Article 3, a nation with a coastline has the right to declare a "territorial sea" outward from its shore up to a limit not exceeding 12 nautical miles.[11] Within this territorial sea the coastal state exercises sovereignty over the water column, the air space above, and the subsoil below.[12] Additionally, coastal states have the exclusive right to conduct marine scientific research within the territorial sea.[13] But this sovereignty is not absolute, because ships of all states enjoy the right of innocent passage through the territorial sea.[14] The nation can regulate the innocent passage of ships within the territorial sea with respect to "the conservation of the living resources of the sea"[15] and "the preservation of the environment of the coastal State and the prevention, reduction and control of pollution thereof."[16]

From 12 nautical miles out to 24 nautical miles away from land is the "contiguous zone." Under Article 33, the coastal state has more limited sovereignty within the contiguous zone, but it may act to prevent infringement of its customs, fiscal, immigration, or sanitary laws and regulations within the contiguous zone.

Next, pursuant to Article 57, the coastal state may elect to declare an exclusive economic zone (EEZ) extending 200 nautical miles beyond the baselines used to determine the territorial seas.[17] In common parlance, the EEZ is often called the "200 mile zone." The declaration of an EEZ allows the coastal state to exercise "sovereign rights" within the EEZ. The nation has the sovereign right to explore, exploit, manage, and conserve the living and nonliving resources of the waters superjacent to the seabed and the seabed and subsoil within the EEZ.[18] The nation thus effectively possesses control over the living resources located within the EEZ, enjoying preferential rights[19] subject to certain limitations.[20] The coastal state is charged with the responsibility of determining the allowable catch of the living resources within the EEZ in order to obtain the maximum sustainable yield, while ensuring that the living resources are not endangered by overexploitation.[21] Again, this provision tries to balance exploitation with sustainability and conservation—a challenging juggling act.

Taken in the aggregate, all of the national EEZs of the planet contain about 30 percent of the world's oceans, about 90 percent of the commercial fisheries, and almost all of the now-exploitable mineral resources.[22] Thus, individual nation-states have the sovereign right, under Article 56, to explore, exploit, conserve, and manage the natural resources (including biodiversity) of a large (although not a majority) portion of the ocean. UNCLOS does attempt to place stewardship obligations on these nations (pursuant to Articles 56, 61, and 62), such as management measures, but the EEZs are largely beyond the reach of international jurisdiction. The nearby nation is mostly in the driver's seat within these significant territories.

While taking these management measures, the coastal state "shall take into consideration the effects on species associated with or dependent upon harvested species with a view to maintaining or restoring populations of such associated or dependent species above levels at which their reproduction may become seriously threatened."[23] The nation also has jurisdiction over scientific research and the protection and preservation of the marine environment within the EEZ.[24] But the coastal state does not enjoy complete sovereignty in the EEZ. Other nations possess the right of overflight, navigation, the laying of cables and pipelines,[25] and access to any surplus of the allowable catch of living resources.[26] Additionally, UNCLOS obligates coastal states to grant, "in normal circumstances," their consent for marine scientific research projects by other states or competent international organizations in their EEZ or on their continental shelf. The state grants its consent with the proviso that the projects be exclusively

for peaceful purposes and in order to increase scientific knowledge of
the marine environment for the benefit of all mankind.[27]

The coastal nation has limited sovereign rights within the continental shelf area.[28] The treaty employs a detailed method for determining the area encompassed within the continental shelf.[29] Under
these procedures the continental shelf can extend out to a distance
between 200 and 350 nautical miles from the baselines used to cal-
culate the territorial sea. The coastal state has no sovereign right over
the superjacent waters or the air space above those waters.[30] The
coastal state has the right to explore and exploit the "mineral and
other non-living resources of the seabed and subsoil together with
living organisms belonging to sedentary species."[31] Sedentary species
are defined as "organisms which, at the harvestable stage, either are
immobile on or under the seabed or are unable to move except in
constant physical contact with the seabed or the subsoil."[32]

Accordingly, the nonsedentary resources and any activity on the
superjacent waters fall under the convention's high-seas regime. The
coastal state is not required to declare a continental shelf, and its
rights are exclusive, meaning that if the state does not explore the
shelf or exploit its resources, no other nation may undertake these
activities without the coastal state's consent.[33] All other nations enjoy
equal rights, with the coastal state, of navigation, overflight,[34] laying
of cables and pipelines,[35] and fishing for nonsedentary species.[36]

Finally, UNCLOS defines the "high seas" as the zone not within
the internal waters, territorial sea, contiguous zone, EEZ, or archipelagic waters of an archipelagic state.[37] In general, the high seas
begin where the EEZ ends, more than 200 nautical miles off the coast.
The high seas are declared by UNCLOS to be *beyond* national jurisdiction, part of the global commons. The convention adopts a legal
regime of freedom of the sea for most activities in this zone,[38] and
no state may subject any portion of the high seas to its sovereignty.[39]
UNCLOS differentiates between living resources and nonliving resources of the high seas. UNCLOS proclaims that the "Area" and its
resources are the "common heritage of mankind."[40] The controversial heart of UNCLOS is this declaration that the deep seabed outside
national jurisdiction is "the common heritage of mankind," with
profits derived from it to be shared by *all* countries. This profit-sharing

provision was the chief reason why the United States failed to ratify. Powerful interests feared that the treaty gave third-world nations too much access to wealth and technologies developed by the United States, and so we refused to agree.[41] It is, after all, difficult to put your money where your treaty is when asked to share no-kidding profits and information gleaned from the common heritage of all people. Lip service is much cheaper.

The "Area" is defined as the "seabed and ocean floor and sub-soil thereof, beyond the limits of national jurisdiction."[42] Under this section of UNCLOS, "resources" are defined as "all solid, liquid or gaseous mineral resources *in situ* in the Area at or beneath the seabed, including polymetallic nodules."[43] All activities for exploration and exploitation of these resources come under the "Authority," and shall be carried out for the benefit of humankind as a whole and for peaceful purposes.[44]

Under the express language of UNCLOS, living resources do *not* fall under the rubric of "common heritage of mankind," but the convention does attempt to impose an obligation on all nations to conserve and manage the living resources of this zone.[45] Contrasted to this, the convention adopts a regime of freedom of fishing for the high seas, but does not define "fishing" or the resources to which this applies.[46] There seems to be some confusion in reconciling these two principles. Presumably, the UNCLOS provision regarding conservation and management applies to all "the living resources of the high seas," including both fishery and nonfishery resources, sedentary and nonsedentary species.[47] However, there is some disagreement with this view. Some have suggested that the convention uses "living resources" only in a fisheries or conservation sense.[48] It is a major failing of UNCLOS that it leaves such a crucial point open to widely different interpretations.

These jurisdictional qualifications, while useful in some respects, splinter the natural resources of the marine environment in some rather arbitrary and unnatural ways. UNCLOS geographically divides areas, in most instances, based on distance from shore, not based on the limits of an ecosystem or other natural boundaries. UNCLOS also divides the resources of a region in ways that do not facilitate cohesive management measures. For example, consider the case of a seamount located within a coastal nation's continental shelf region. The coastal nation would be able to exercise control over the sea floor, minerals, and sedentary species of that seamount, but not the living resources swimming above and around that same spot.

Likewise, imagine a seamount that sits directly upon the 200-nautical-mile line. The coastal state could exercise preferential control over the nonsedentary living resources in the superjacent water within the 200-nautical-mile limit, but would have no control over these same resources after crossing this artificial boundary. It is under this fractured jurisdictional framework that we must examine the existing legal treaties and laws regarding their effectiveness in protecting the biodiversity of these seamount and vent communities and other marine hotspots.

As demonstrated in chapter one, ocean jurisdiction has developed into a zonal system, and thus, the location of a seamount, coral reef, or hydrothermal vent will determine under which jurisdictional regime it falls. In the natural world, of course, seamounts, reefs, hydrothermal vents, and their associated species do not conform to these artificial boundaries. Such artificial lines have no meaning in the natural world. Thus, a coral reef, seamount, or vent area could be covered under a coastal nation's domestic legislation, several nations' domestic legislation, a coastal nation's regional or international treaty obligations, customary international law, or any combination of these.

While this piecemeal and overlapping structure leaves a lot to be desired, it would seem rational for international organizations and individual nations to develop comprehensive management schemes for the entire area under their control. But, to date, the attempts to manage the oceans have typically been piecemeal, focusing not on ecosystems as a whole or on cumulative impacts, but rather on particular ocean uses.[49] Additionally, prior to UNCLOS (1982), the international community did not comprehensively and directly address the issue of marine conservation.[50] Customary international law, such as it is, was about all that was available in terms of legal structure.

All nations are expected to abide by customary international law, although these expectations are often frustrated. Customary international law is thought to be generated gradually and informally through common state practice and *opinio juris*, a "sense of legal obligation" for states to follow a certain practice.[51] If enough nations act like something is an international legal principle, and do so for a long enough time, at some vague and nebulously defined point it may become "the law." Several doctrines that relate to the protection of biodiversity of areas within a nation's control may be considered binding customary law. Among these are the "precautionary principle" and Principle 21 of the Stockholm Declaration.

The precautionary principle generally states that "in order to protect the environment, the precautionary approach shall be widely applied by States according to their capabilities. Where there are threats of serious or irreversible damage, lack of full scientific certainty shall not be used as a reason for postponing cost-effective measures to prevent environmental degradation."[52] In other words, if a matter is vital and if an error could be damaging and permanent enough to the environment, nations should put into effect the "better-safe-than-sorry" philosophy and not wait around for definitive scientific confirmation of the crisis. The precautionary principle is considered by some to have attained the status of customary international law,[53] and is embodied in several environmental treaties.[54] But this opinion is open to considerable debate.[55]

A related precept of customary international law is Principle 21 of the Stockholm Declaration.[56] Principle 21 provides that "states have . . . the sovereign right to exploit their own resources pursuant to their own environmental policies, and the responsibility to ensure that activities within their jurisdiction or control do not cause damage to the environment of other states or of areas beyond the limits of national jurisdiction."[57] Principle 21 was included, in slightly varied form, in the Rio Declaration.[58] It is important to note that Principle 21 makes states responsible for actions within their control which cause damage to the environment of areas *beyond* the limits of national jurisdiction. This clearly indicates that states are expected to control and regulate the activities of ships flying their flag and has important implications for protecting the high-seas environment, as will be discussed below.

Because the precautionary principle does not contain any jurisdictional limitation, it would seem to require states to relinquish the short-term financial opportunities available from resource depletion and loss of biodiversity in order to protect long-term human benefits for the planet. In the marine context, these tenets readily apply to the territorial sea, where the coastal state manifests sovereignty. The coastal state is free to manage its environment in the way that it sees fit, subject to binding and enforceable international obligations. Similarly, under UNCLOS, in the EEZ and continental shelf areas, the coastal state does not possess sovereignty, merely sovereign rights. Consequently, the rights and responsibilities of the coastal state are granted by the convention (and enforcement would be through the convention's procedures).[59]

UNCLOS expressly grants nations the sovereign right to exploit their natural resources pursuant to their environmental policies and in accordance with their duty to protect and preserve the marine environment.[60] Furthermore, UNCLOS does not attach any spatial differentiation to this definition of natural resources, so presumably this right exists in any area where UNCLOS grants the coastal state control over the natural resources (i.e., the EEZ and continental shelf).

These principles have several shortcomings. First, the aforementioned precautionary principle contains the devastating caveat "according to their capabilities." Although UNCLOS states in Article 194 that parties "shall" take all measures necessary to prevent, reduce, and control marine pollution from any source, this seemingly strong provision is immediately weakened in Article 194(1) with the proviso that parties shall make these efforts "in accordance with their abilities." This caveat has caused some analysts to conclude that the precautionary principle may be merely hortatory language that is intended to guide states as they adopt national legislation and plans.[61] The inclusion of this caveat in such a key provision has had roughly the same effect one might imagine if the federal tax code stated that every taxpayer must comply with all applicable obligations to pay his or her income tax in full every year, according to his or her capabilities. It would not take long before the federal government would need to hold a bake sale to keep the interstate highway system repaired.

This "permissive approach" to resource use and human activity creates a complex balancing of interests that makes it possible for developmental and quality of (human) life considerations to outweigh the need to conserve biodiversity and to take other environmentally oriented preventive action.[62] Thus, states may claim that they are unable to comply with their supposed conservation duties due to their economic, food, or resource needs. Although an expansive application of the precautionary principle may someday come about, the permissive interpretation dominates the status quo today.[63] Just as with bringing up children, a permissive approach to the law of the sea guarantees spoiling. It is all too predictable that nations often discover that other pressing needs prevent them, much to their dismay, from doing anything to preserve biodiversity in the oceans. Even a mass extinction cannot force countries to help bail out the leaking Ark when they can instead devote themselves to whining that their own lifeboats need patching.

Additionally, there is a problem with enforcement of conservation responsibilities. Unless some transboundary damage is implicated, no state may raise a legal objection to the domestic environmental policies of any other state.[64] At present, violation of the precautionary principle does not constitute a breach of international law.[65] Within a state's own borders, international law permits the state to deplete or injure its natural resources, to destroy its gene pool, species, and habitats, and to otherwise harm its environment.[66] Whatever its other virtues, national sovereignty is a very effective cover for nations inclined to exploit "their" marine resources for all they are worth, consequences be damned.

These principles essentially create a framework similar to that of the United States' National Environmental Policy Act (NEPA).[67] NEPA requires the consideration of environmental effects in decision making, and requires the preparation of detailed environmental impact statements that must take a hard look at the foreseeable consequences of the federal action.[68] However, the end result is similar to that achieved under the combination of the precautionary principle and Principle 21; the ultimate decision maker does not have to follow the most environmentally favorable choice, but instead must only go through the process of considering options and examining the consequences of a given action.[69] The international community must simply stand by and is not empowered to substitute its judgment for that of the individual nation.[70]

Consistent with the high-seas regime of UNCLOS, a nation cannot directly apply the precautionary principle to areas not under its control. However, a nation has the "duty to take, or to cooperate with other States in taking, such measures for their respective nationals as may be necessary for the conservation of the living resources of the high seas."[71] Thus, at least in theory, a nation, applying the precautionary principle to its fullest extent, has an obligation to ensure that its own citizens and corporations do not engage in activities that could cause irreversible harm to living resources. If applied by all nations, this potentially could be a powerful rule to enact, at least indirectly, for marine protected areas on the high seas.[72] But any extension of a nation's environmental policies beyond its own territory would need to comply with the General Agreement on Tariffs and Trade (GATT) and avoid imposing negative and discriminatory restrictions on the free trade of other countries on the oceans.[73] This poses some formidable obstacles for the traditional sanctions-based approach predicated on environmental trade measures.

Notwithstanding this unenforceable and nebulous standard of UNCLOS, some nations have seized its philosophy and applied the idealization of the precautionary principle in their domestic environmental policy to establish marine protected areas (MPAs). I will return to this topic later in the book, but it is necessary to establish some baseline concepts at this juncture in order to understand some key strengths and weaknesses of UNCLOS and other international agreements as applied to marine biodiversity.

It is highly significant that UNCLOS, the King Neptune of all marine laws, does *not* explicitly and clearly mandate the creation and maintenance of a well-chosen worldwide network of MPAs. These marine enclaves could be the key to halting and reversing the mass extinction now underway. Even without much help from UNCLOS, there are approximately 1,300 MPAs in the world today, but the lack of overarching legal structure has led to predictable problems.[74] Marine protected areas come in many forms and incorporate a variety of restrictive measures. Some are true sanctuaries, theoretically prohibiting all activities that may be harmful to the protected area. Unfortunately, many others are more limited in scope, prohibiting only certain fishing practices or commercial shipping. A study concluded in 1995 that, of the 383 MPAs assessed, only 31 percent were generally achieving their management objectives.[75]

One major problem is the same flaw that is often found in terrestrial parks, preserves, and refuges: In the legal vacuum left by UNCLOS, nations have chosen to protect that which does not need it, and failed to protect that which does. In fact, there is evidence that most MPAs are not optimally sited and are too small to safeguard the marine biodiversity within them, and that many globally unique marine habitats are not covered by any MPA.[76]

Marine protected areas, if properly selected and protected, can provide several benefits to an ecosystem and its human populace. Fisheries benefit from enhanced fecundity and the spillover of adult and juvenile fish into nonprotected fishing grounds.[77] The local economies benefit from ecotourism and the beneficial effect of the above spillover.[78] The ocean benefits from the protection of habitat, or the recovery of degraded habitat, and the existence of a more natural ecosystem in which adult and top predator fish exist.[79] Also, humanity benefits from the management and protection of biodiversity within these areas.[80] Finally, MPAs can simplify resource management by substituting clear restrictions for some of the complex rules presently employed in most fisheries (such as what can and cannot be caught, and when and where fishing activity can proceed).[81]

It is estimated that less than half of one percent of the world's oceans are protected by MPAs, and some oceans are protected in name only.[82] Furthermore, almost 80 percent of these preserves are not actively managed at all and thus exist only on paper.[83] In general, these ostensibly protected areas fail due to lack of local commitment. The local communities fail to see or appreciate any local benefits from the protected areas, and thus are unwilling to embrace the creation of protected areas or to comply with the restrictions.[84] Additionally, these protected areas are only as protected as the areas around them. If areas are improperly sized or situated, activities such as intensive farming, fishing, mining, or timber cutting adjacent to the MPA can effectively negate any benefits of the restrictions.[85]

Furthermore, developing nations may have a strong incentive not to establish MPAs in resource-rich areas. Many developing nations are not in a position to afford to bypass the mineral, energy, and food resources of their marine environment. In addition, developing nations lack the technology and scientific resources to identify these relatively small and unique features within their coastal environment. As UN-CLOS grants the coastal state the right to consent to, and limit the extent of, marine scientific research within its territorial sea and EEZ, some nations may obstruct researchers attempting to locate these areas. A state can resist international pressure to protect these areas by "validly" claiming that unique coral reef, seamount, or hydrothermal vent areas do not exist within their coastal environment.

Notwithstanding these problems, several states are attempting to protect these areas.[86] The United States is considering designating the Davidson Seamount, off the California coast, as part of the Monterey Bay National Marine Sanctuary.[87] The Australian government has recently taken steps to protect seamounts off the coast of Tasmania.[88] But, simply designating isolated areas as "protected," is meaningless without significant restrictions, ample buffer zones, and stringent enforcement.[89]

Finally, and most importantly, there currently exists *no* express legal authority (in UNCLOS or elsewhere) for designating MPAs in the crucially important high-seas region.[90] This is the prime example of the bundle of missed opportunities we call UNCLOS. Yet MPAs went drifting by while the focus of UNCLOS was firmly riveted to economic development and exploitation rather than preservation of marine natural resources. Consequently, immense and vital areas are not within direct legal protection as any type of international reserve, despite the evident ecological need.[91] This is the same monumental flaw that vitiates much of the promise of another major international law, the World Heritage Convention, which I will discuss shortly. Incomprehensibly vast areas of the open ocean are entirely beyond the reach of the Law of the Sea Treaty insofar as MPAs are concerned; thus, the best chance to safeguard marine hotspots in these areas has been missed by international law. The UNCLOS ship has sailed, and MPAs have been left stranded.

Let us return briefly to the niche that biodiversity issues occupy within the vast ambit of UNCLOS. UNCLOS is intended, theoretically at least, to promote a legal program covering *all* uses of the oceans, and to be conducive to the equitable and efficient use of resources, the conservation of living resources, and the protection and preservation of the marine environment.[92] In addition to establishing the sovereign rights and ocean boundaries described above, UNCLOS attempts to set down some obligations on member states with respect to the conservation and utilization of living resources and the protection of the marine environment.

All states have the nice-sounding but content-challenged "obligation to protect and preserve the marine environment."[93] The convention requires that states "shall cooperate on a global basis and, as appropriate, on a regional basis, directly or through competent international organizations, in formulating and elaborating international rules, standards and recommended practices and procedures consistent with this Convention, for the protection and preservation of the marine environment, taking into account characteristic regional features."[94] This provision is vague and contains no substantive provisions relating to protecting unique areas of biodiversity. A nation only needs to "cooperate," in forming rules that are consistent with the convention; the convention does not truly and explicitly address biodiversity at all.

UNCLOS specifically requires that member states take actions "necessary to protect and preserve rare or fragile ecosystems as well as the habitat of depleted, threatened or endangered species and other forms of marine life."[95] While this sounds promising for reef, vent, seamount, and other hotspot communities, this section is expressly limited to Article 194 (dealing only with pollution of the marine environment) and does *not* obligate states to prevent exploitation or other threats to these rare ecosystems. Thus, its utility is confined to allowing a coastal state to enact stricter pollution control standards in the vicinity of rare or fragile ecosystems located within its territorial sea or EEZ.

UNCLOS also requires states to conserve and manage the living resources in the EEZ, with the oxymoronic goal of "optimal utilization" of these resources.[96] While managing these resources, the coastal state "shall take into consideration the effects on species associated with or dependent upon harvested species."[97] Thus, coastal states have an affirmative duty to employ a form of ecosystem management for the living resources of the EEZ, and to ensure that while managing exploitable resources, they consider the effect of removing especially important species such as keystone species. But, aside from combining conservation with exploitation (which, unlike mixing apples and oranges, is more akin to mixing pineapples and hand grenades), this provision lacks teeth.

Reminiscent of NEPA, the coastal state is only required to "consider" the effect on associated species. There is nothing that would substantively obligate a coastal state to refrain from exploiting a species on which another species of a coral reef, seamount, or vent community were dependent. Likewise, the convention does not require that coastal nations employ the precautionary principle. Given the woefully undeveloped state of the world's knowledge about marine species and their interactions, this is a major omission on the part of UNCLOS.

Additionally, states are required to take action to prevent pollution of the sea caused by activities on land, in the sea, and in the atmosphere. States are required to take "all measures consistent with this Convention that are necessary to prevent, reduce and control pollution of the marine environment from any source, using for this purpose the best practicable means at their disposal."[98] But this obligation is qualified by the following "get-out-of-jail-free" clause which we have seen before, "in accordance with their capabilities."[99] Furthermore, coastal states are allowed to dump waste, and to permit other nations to dump waste, within their territorial sea and EEZ.[100]

The convention employs several mechanisms in an inchoate effort to put teeth into these exception-encrusted environmental protection obligations. UNCLOS declares that nations are "responsible for the fulfillment of their international obligations concerning the protection and preservation of the marine environment" and that they "shall be liable in accordance with international law."[101] UNCLOS provides for binding dispute resolution in certain situations, including "when it is alleged that a coastal State has acted in contravention of specified international rules and standards for the protection and preservation of the marine environment which are applicable to the coastal State and which have been established by this Convention or through a competent international organization or diplomatic conference in accordance with this Convention."[102] Finally, nations are only allowed to make declarations or statements regarding the convention's application at the time of signing, ratifying, or acceding to UNCLOS, which does not purport to exclude or modify the legal effect of the provisions of the convention.[103]

While there are several positive features in UNCLOS, the convention misses the boat in some important respects. First, as I have mentioned, the treaty establishes separate zones in which the coastal state possesses a varied bundle of legal rights, allowing for the coastal state to manage a variety of activities in these zones. But these zones do not conform to the natural environment, and instead are unnatural arbitrary lines. While advantageous from a practical standpoint, these lines do not allow for meaningful ecosystem-oriented governance. The biological and physical realities are such that there are significant interactions between the waters, atmosphere, land, and life forms of a given area, and the treaty sidesteps these interactions in many circumstances.[104] These artificial boundaries and the resulting disparate treatment of adjacent resources do not account for the physical realities of the highly complex interactions that occur in the marine environment.

While superficially addressing the need for regional and international cooperation in managing regions and resources, these goals seem largely aspirational, a set of lofty goals to be aimed at, or wished for. The convention obligates coastal states to enact regulations protecting and preserving the marine environment, but does not contain many substantive provisions, even in key subjects such as marine protected areas, leaving many of these details to be worked out through international or regional agreements, or by the coastal nation itself... if at all. Similarly, UNCLOS has a decidedly prodevelopment stance, focusing most of its efforts on establishing coastal nations' rights to develop, harvest, and exploit the living and nonliving resources of the regions. To say the least, this economic orientation deflects attention away from preserving biodiversity, and it may often generate activities affirmatively harmful to marine life. No one can serve two masters well, and UNCLOS is clearly aimed at favoring Greenback, the god of money. Conservation is lucky to have so much as a steerage ticket on the UNCLOS cruise ship.

One last specific situation deserves mention within the subject of UNCLOS. Under the authority of UNCLOS, a regime has been established to manage so-called straddling stocks and highly migratory fish. The United Nations Agreement for the Implementation of the Provisions of the United Nations Convention on the Law of the Sea relating to the Conservation and Management of Straddling Fish Stocks and Highly Migratory Fish Stocks (Agreement on Straddling Stocks)[105] sets out principles for the conservation and management of those boundary-crossing fish stocks and establishes that such management must be based on the precautionary approach[106] and the best available scientific information.[107] While primarily concerned with ensuring the optimum utilization of fisheries resources,[108] the Agreement on Straddling Stocks does say that coastal states and states fishing on the high seas shall "adopt, where necessary, conservation and management measures for species belonging to the same ecosystem, associated with, or dependent upon the target stocks, with a view to maintaining or restoring populations of such species above levels at which their reproduction may become seriously threatened."[109]

Along with this somewhat nebulous edict to apply ecosystem management, the agreement also vaguely asserts that coastal states and states fishing on the high seas shall "protect biodiversity in the marine environment."[110] This provision is notably a step forward in that it tries to obligate states to protect biodiversity in the marine environment without distinguishing between areas of jurisdiction or between fishery resources and nonfishery resources. But in context, this biodiversity goal lacks any substantive standards and seems largely aspirational, inasmuch as the rest of the agreement focuses on establishing regional pacts and setting standards for the exploitation, conservation, and management of commercially valuable straddling fish stocks and highly migratory fish stocks.

To sum up, UNCLOS does *not* address biodiversity conservation comprehensively, explicitly, and directly, and accordingly fails to protect the enormous variability of marine species or ecosystems. It is a work of obvious compromise and political maneuvering, and displays telltale signs of trying to serve too many irreconcilable interests. The convention's pronounced emphasis on fostering free trade and exploitation of the oceans by all nations for economic gain does not fit well with the few bones thrown to conservationists. This inadequacy is particularly prominent in areas outside of national jurisdiction; traditional notions of freedom of the seas are invoked, perhaps incorrectly, by parties exploring and exploiting high-seas resources who proclaim that their actions are subject to no laws or regulations other than those dictated or agreed to by the flag state.[111] Where UNCLOS might have furthered conservation there is instead consumption, and where it might have stressed nature preserves it instead preserves stress on nature. UNCLOS might have established freedom for biodiversity in the high seas, but the only freedom it embraces is the license of self-interest. Thus, UNCLOS, the one international agreement theoretically most suited for marine biodiversity protection, is at best only a very incomplete response to the mass extinction bubbling under the surface of the planet's waters.

CONVENTION ON BIOLOGICAL DIVERSITY

At the 1992 United Nations Conference on Environment and Development (the "Earth Summit") in Rio de Janeiro, leaders of many nations gathered to discuss "sustainable development." One product of that meeting was the Convention on Biological Diversity (CBD).[112] Over 150 governments signed the document at the Rio conference, and since then a total of 188 countries have become parties to the agreement.[113]

The CBD establishes three main goals: the conservation of biological diversity, the sustainable use of its components, and the fair and equitable sharing of the benefits from the use of genetic resources.[114] In contrast to earlier treaties, it does not include any lists or annexes of protected species or areas, but deals with the problem of biodiversity in a more comprehensive fashion, addressing all aspects of biodiversity including access to biological resources, biotechnology, and financial resources.[115] The CBD identifies the problem of dwindling biodiversity, sets overall goals and policies and general obligations, and organizes technical and financial cooperation. However, the responsibility and discretion for achieving its goals rests largely with the countries that sign and ratify it.

For example, Article 6 states that signatories are required, in accordance with their "particular conditions and capabilities," to "develop national strategies, plans or programmes for the conservation and sustainable use of biodiversity," and "as far as possible and as appropriate, [to integrate] the conservation and sustainable use of biological diversity" into broader national plans for environment and development.[116] This is a laudable idea, but the caveats and conditional clauses weaken it considerably. Nations are left to decide for themselves whether they have the "conditions and capabilities" that generate these duties, and whether any given actions are "possible" or "appropriate."

Similarly, Article 7 provides that each signatory shall "as far as possible and as appropriate" "identify components of biological diversity important for its conservation and sustainable use" and monitor them, "paying particular attention to those requiring urgent conservation measures and those which offer the greatest potential for sustainable use."[117] In identifying these key areas, nations are to consider, among other things, ecosystems and habitats

> containing high diversity, large numbers of endemic or threatened species, or wilderness; required by migratory species; of social, economic, cultural or scientific importance; or, which are representative, unique or associated with key evolutionary or other biological processes.[118]

This description certainly is broad enough to embrace both the marine and terrestrial biodiversity hotspots, if a nation is so inclined. But when is it "possible" for a developing nation to divert scarce resources to the identification and monitoring of key pockets of biodiversity? When is it "appropriate" to make such an investment in biodiversity, in light of all the other pressing needs that poorer nations must try to meet (or that even wealthy nations must meet, for that matter)? When the nation itself decides, it is not surprising that the default answer fluctuates wildly between seldom and never.

Article 8, pertaining to in-situ conservation, includes the same escape hatch—"as far as possible and as appropriate"—in directing each signatory, inter alia, to establish "a system of protected areas or areas where special measures need to be taken to conserve biological diversity"; to develop guidelines for their selection, establishment, and management; to "[r]egulate or manage biological resources important for the conservation of biological diversity whether within or outside protected areas, with a view to ensuring their conservation and sustainable use"; to "[p]romote the protection of ecosystems, natural habitats, and the maintenance of viable populations of species in natural surroundings"; to "[p]romote environmentally sound and sustainable development in areas adjacent to protected areas with a view to furthering protection of these areas"; and to "[r]ehabilitate and restore degraded ecosystems and promote the recovery of threatened species . . . through the development and implementation of plans or other management strategies."[119]

The concept of sustainable use appears again in Article 10. Once more, "as far as possible and appropriate," signatories are to, inter alia, "[i]ntegrate consideration of the conservation and sustainable use of biological resources into national decision-making," "[a]dopt measures relating to the use of biological resources to avoid or minimize adverse impacts on biological diversity," encourage cooperation between government and the private sector in developing methods for sustainable use, and "[p]rotect and encourage customary use of biological resources in accordance with traditional cultural practices that are compatible with conservation or sustainable use requirements."[120]

Its obvious shortcomings aside, the CBD was a breakthrough in a way because the text of Stockholm Principle 21 appears verbatim as Article 3, marking the first time this language had appeared in binding international law, rather than in "customary law" or "soft law."[121] Article 3 reads:

> States have, in accordance with the Charter of the United Nations and the principles of international law, the sovereign right to exploit their own resources pursuant to their own environmental policies, and the responsibility to ensure that activities within their jurisdiction or control do not cause damage to the environment of other States or of areas beyond the limits of national jurisdiction.[122]

Thus, the traditional concept of national sovereignty over resources is, at least in principle, balanced within the CBD by the
requirement that each party accept its responsibility not to harm
the territory of any other state or the territory beyond its own national jurisdiction.[123] This could be significant for marine and terrestrial hotspots preservation, given the global significance of these
eco-regions and the persistent problem of individual nations exploiting them and/or failing to afford them adequate protection. At
present this is mostly an unrealized potential, but such is the sorry
record of achievement in international law that even the mere mention of this concept in the text of a treaty is an unusual milestone.

The CBD also contains a progressive provision in terms of
funding. In recognition of the practical concerns and needs of many
countries, CBD-related activities in developing countries are eligible
for support from the financial mechanism of the CBD (i.e., the Global
Environment Facility [GEF]).[124] Each party is to provide financial
support according to its available resources and commensurate with
the national objectives undertaken to meet the CBD's directives.[125]
GEF projects, supported by the United Nations Environment Program (UNEP), the United Nations Development Program (UNDP)
and the World Bank, are to help forge international cooperation and
finance actions to address four critical threats to the global environment: biodiversity loss, climate change, depletion of the ozone layer,
and degradation of international waters. By the end of 1999, the GEF
had contributed nearly $1 billion for biodiversity projects in more than
120 countries.[126] Undoubtedly, many of these projects have made a
valuable contribution to the cause of biodiversity.

This is a promising feature of the CBD. It acknowledges the need
to provide positive financial incentives for biodiversity conservation,
which contrasts with the traditional fear-driven command-and-control
model.[127] It also recognizes that some nations are more capable
than others of funding environmental protection and seeks to level
the playing field. Without a meaningful infusion of resources, including and especially money, many developing nations will lack
the wherewithal to effect real progress within their borders, no matter
how devoted they may be to the theoretical ideal of environmental
protection.

On a related topic, one of the most important portions of the CBD is the provision for technical and scientific cooperation and the creation of a mechanism to collect, manage, and disperse information and statistics on global biodiversity.[128] Article 18 provides for a clearinghouse whereby technical and scientific information concerning biodiversity can be shared among nations to help further conservation efforts.[129] This, too, is a worthwhile initiative that could globally advance the state of knowledge in the area where it is most desperately needed—the world's ocean waters. Developing nations might get a head start on their own conservation programs if they can build on the lessons learned by other, more resource-rich, developed countries.

The CBD is a broad document which also features some very controversial provisions on intellectual property and biotechnology, including "genetic resources." For example, Article 19 provides in part:

> Each Contracting Party shall take all practicable measures to promote and advance priority access on a fair and equitable basis by Contracting Parties, especially developing countries, to the results and benefits arising from biotechnologies based upon genetic resources provided by those Contracting Parties. Such access shall be on mutually agreed terms.[130]

Similarly, Article 16 provides for access to and transfer of technology, including biotechnology, such that developing countries "which provide genetic resources are provided access to and transfer of technology which makes use of those resources, on mutually agreed terms, including technology protected by patents and other intellectual property rights."[131] Such terms have been, not surprisingly, of great concern to the United States, to the extent that President Clinton did not even seek Senate ratification of the CBD. Developing nations may claim that financial profits from the exploitation and development of natural resources, along with highly valuable and expensively acquired intellectual property (including biotechnology), are being siphoned off by other nations. This specter has haunted the CBD and has frightened away the United States and some other developed nations.[132] This, of course, is very similar to the situation under UNCLOS, where the United States has remained on the sidelines. It is one thing to sign on to a pretty-sounding collection of hopes and dreams, but it is quite another to agree to share valuable information and even profits with the nations that have not "earned" them.

The United States' prolonged bout of hesitancy notwithstanding, many key terrestrial- and marine-hotspot nations have signed and ratified or otherwise approved the CBD, including Brazil, Madagascar, Papua New Guinea, Democratic Republic of the Congo, China, India, Indonesia, and others.[133] In theory, then, the CBD could be a useful tool for hotspots preservation. However, there are some serious shortcomings, as we have already begun to see.

First, the CBD does not actually create enforceable legal obligations. Instead, it directs its signatories to enact legislation within their jurisdiction, consistent with the CBD objectives. If a nation fails to do this, there are no real consequences. Under the CBD, parties have very few obligations, and most of these are eviscerated with the gaping loophole phrases "as far as possible and as appropriate," or "in accordance with [a party's] capabilities."[134] For developing nations, implementing measures are further contingent on commitments from first-world parties to provide technology and funding.[135] In addition, no mechanism exists to assess the substantive adequacy and consistency of national biodiversity plans, and thus it is practically impossible to detect any breach of CBD obligations.[136] Where standards are so vague, self-defining, and gap-toothed, no one can tell when they are being violated.

The GEF has been criticized as well. Some have noted the conflict inherent in the involvement of the World Bank as the managing partner of the GEF. The World Bank might impose a prodevelopment inclination on CBD actions, reminiscent of UNCLOS.[137] Additionally, critics have pointed out a possible GEF/World Bank bias toward supporting projects that redound to the benefit of developed nations instead of devoting more attention to developing countries, where most of the real action should be.[138]

Also, the CBD does not focus on hotspots per se, but only instructs signatories to identify and monitor important biodiversity resources and take some steps toward preserving them. This is very general guidance, and compliance is very much in the eye of the beholder.[139] There is no overarching priority scheme for either identifying or protecting the most vital pockets of biodiversity. Worthwhile initiatives may be fostered by the CBD, nation by nation, one park or preserve at a time, but this is too haphazard and idiosyncratic to be a substitute for specific, big-picture legislation with real enforcement capabilities. It is fine that the CBD at least allows its signatories to focus attention on marine hotspots for conservation priorities, but resources of this importance deserve more express and affirmative support.

To summarize, the CBD has three explicit aspirations: the conservation of biological diversity, the sustainable use of its components, and the fair and equitable sharing of the benefits from the utilization of genetic resources.[140] The CBD requires states to monitor, study, and catalogue the genetic resources contained in their rain forests, coral reefs, wetlands, deserts, and coastal zones. The CBD contains the highly controversial provision that developing nations which "provide genetic resources are provided access to and transfer of technology which makes use of those resources, on mutually agreed terms, including technology protected by patents and other intellectual property rights."[141] Many developed nations, including the United States, object to these provisions and have refused to ratify the convention.[142] Furthermore, the CBD fails to adequately address access to and ownership rights of the genetic resources of areas outside national jurisdiction.

The CBD codifies Principle 21, and clearly establishes that states have the sovereign right to exploit their own resources located within their jurisdiction (including marine resources) pursuant to their own environmental policies. States also have the responsibility to ensure that activities within their jurisdiction or control do not cause damage to the environment of other states or of areas beyond the limits of national jurisdiction.[143] While laudable in establishing that each party has the responsibility to prevent harm to areas outside its own national jurisdiction, the CBD does little to ensure that parties in fact preserve the resources located entirely within their jurisdiction, including the coastal waters.

Additionally, each state shall, "as far as possible and as appropriate," cooperate with other contracting parties regarding conservation and sustainable use of biological diversity in areas beyond national jurisdiction.[144] This ambiguous obligation to "cooperate," is further weakened by the fishy quibble "as far as possible and as appropriate." The CBD expressly states that "Contracting Parties shall implement this Convention with respect to the marine environment consistently with the rights and obligations of States under the law of the sea."[145] Thus, regarding the high seas, a nation's exercise of its right of freedom of the seas that threatens or harms biodiversity would take precedence over the conservation and sustainable-use obligations of the CBD.[146] This is the same way the scales tip toward the dollar under UNCLOS. When push comes to shove, the way is clear to push for profits and tell conservationists to shove it.

The CBD tantalizes and teases conservationists with Article 8(a), which charges contracting parties with the obligation to "as far as possible and as appropriate ... [e]stablish a system of protected areas or areas where special measures need to be taken to conserve biological diversity."[147] But, in concert with Articles 4 and 22(2) (and, by provision, UNCLOS) individual contracting parties are ostensibly prohibited from unilaterally establishing a system of protected reserves in areas *beyond* national jurisdiction.[148] Lastly, the CBD fails to provide incentives to states to protect marine biodiversity beyond the continental shelf and EEZ. Consequently, vast areas of the world's biosphere are left inadequately protected.

In 1995 the CBD conference of the parties recognized that there were gaps and deficiencies in the CBD structure and requested the Executive Secretary of the Subsidiary Body of Scientific, Technical and Technological Action (SBSTTA), in consultation with the U.N. Office for Ocean Affairs, to undertake a study of the relationship between the CBD and UNCLOS with regard to the conservation and sustainable use of genetic resources of the deep seabed.[149] The SBSTTA suggested four alternative ways of regulating marine biodiversity beyond the continental shelf and EEZ: (1) preserve the status quo, (2) amend UNCLOS, (3) amend the CBD, and (4) negotiate a new regime.[150] Clearly, the status quo in terms of marine biodiversity was perceived by these experts as badly in need of attention. The world is still waiting to see what solution they might produce.

WORLD HERITAGE CONVENTION

The Convention Concerning the Protection of the World Cultural and Natural Heritage (the World Heritage Convention, or WHC)[151] was adopted by the General Conference of the United Nations Educational, Scientific, and Cultural Organization (UNESCO) in 1972. The WHC provides an international framework for the protection of natural and cultural areas of "outstanding universal value."[152] To date, some 177 countries have adhered to the WHC (the overwhelming majority of the member states of the United Nations), including key nations with both terrestrial and marine hotspots.[153]

The preamble states with clarity the core principles relevant to the preservation of all resources that are locally situated yet have global significance. Although neither the term "biodiversity hotspot," nor any of the alternative means for establishing biodiversity conservation priorities (e.g., Global 200 Ecoregions, Endemic Bird Areas, Centres of Plant Diversity, WORLDMAP)[154] specifically appear anywhere in the WHC, the vexing challenges that assail such natural treasures are nonetheless recognized in the preamble:

[T]he cultural heritage and the natural heritage are increasingly threatened with destruction not only by the traditional causes of decay, but also by changing social and economic conditions which aggravate the situation with even more formidable phenomena of damage or destruction. . . . [D]eterioration or disappearance of any item of the cultural or natural heritage constitutes a harmful impoverishment of the heritage of all the nations of the world. . . . [P]rotection of this heritage at the national level often remains incomplete because of the scale of the resources which it requires and of the insufficient economic, scientific, and technological resources of the country where the property to be protected is situated. . . . [E]xisting international conventions, recommendations and resolutions concerning cultural and natural property demonstrate the importance, for all the peoples of the world, of safeguarding this unique and irreplaceable property, to whatever people it may belong. . . . [P]arts of the cultural or natural heritage are of outstanding interest and therefore need to be preserved as part of the world heritage of mankind as a whole. . . . [I]n view of the magnitude and gravity of the new dangers threatening them, it is incumbent on the international community as a whole to participate in the protection of the cultural and natural heritage of outstanding universal value, by the granting of collective assistance which, although not taking the place of action by the State concerned, will serve as an efficient complement thereto...[and] it is essential for this purpose to adopt new provisions in the form of a convention establishing an effective system of collective protection of the cultural and natural heritage of outstanding universal value, organized on a permanent basis and in accordance with modern scientific methods.[155]

Building on this philosophical and factual predicate, the WHC establishes, as its centerpiece, a list of specific places in the world that meet its overarching criterion of "of outstanding universal value." The World Heritage List is the compendium of sites, in either the "natural heritage"[156] or "cultural heritage"[157] category, that have been recognized formally according to the terms of the WHC.

The WHC defines the type of natural or cultural sites that can be considered for inclusion in the World Heritage List, and sets forth the duties of states parties in identifying potential sites and in protecting them. Specifically with regard to "natural heritage" sites, the WHC supplies the following criteria:

[N]atural features consisting of physical and biological formations or groups of such formations, which are of outstanding universal value from the aesthetic or scientific point of view; geological and physiographical formations and precisely delineated areas which constitute the habitat of threatened species of animals and plants of outstanding universal value from the point of view of science or conservation; natural sites or precisely delineated natural areas of outstanding universal value from the point of view of science, conservation, or natural beauty.[158]

In Article 4 the convention places the primary "duty of ensuring the identification, protection, conservation, presentation and transmission to future generations of the cultural and natural heritage" sites on the World Heritage List with the nation that is host to each site.[159] Each host nation is to "do all it can to this end, to the utmost of its own resources."[160] Additionally, where appropriate, each host nation may also draw upon "any international assistance and co-operation, in particular, financial, artistic, scientific and technical, which it may be able to obtain."[161] More detailed requirements are delineated in Article 5, which unfortunately prefaces its worthy mandates with the multilayered qualifier that each state party "shall endeavor, in so far as possible, and as appropriate for each country":

(a) to adopt a general policy which aims to give the cultural and natural heritage a function in the life of the community and to integrate the protection of that heritage into comprehensive planning programmes; (b) to set up within its territories, where such services do not exist, one or more services for the protection, conservation and presentation of the cultural and natural heritage with an appropriate staff and possessing the means to discharge their functions; (c) to develop scientific and technical studies and research and to work out such operating methods as will make the State capable of counteracting the dangers that threaten its cultural or natural heritage; (d) to take the appropriate legal, scientific, technical, administrative and financial measures necessary for the identification, protection, conservation, presentation and rehabilitation of this heritage; and (e) to foster the establishment or development of national or regional centres for training in the protection, conservation and presentation of the cultural and natural heritage and to encourage scientific research in this field.[162]

This is an ambitious agenda, but one rendered hostage to the whims of the leadership within each state party. Nations that are predisposed to take effective action to preserve their natural and cultural heritage will do so, and probably would do so even without Article 5 of the WHC. Those that lack this predisposition will find ample room for discretion and exception in the introductory clause to justify a very comfortable inaction. As a result, the efficacy of these provisions is questionable even within the confines of Article 5 itself. Other more overarching problems with the WHC have further impaired the convention in its implementation and enforcement, as will be discussed shortly.

Article 6 is at the core of the WHC, insofar as it is a potential source of succor for at least some of the hotspots of the world, because it declares that the World Heritage List sites are indeed a world heritage, which the entire international community has a duty to protect in a cooperative effort. But, as with Article 5, it also begins with an important caveat:

> Whilst fully respecting the sovereignty of the States on whose territory the cultural and natural heritage . . . is situated, and without prejudice to property rights provided by national legislation, the States Parties to this Convention recognize that such heritage constitutes a world heritage for whose protection it is the duty of the international community as a whole to cooperate.[163]

Article 6 provides further details, including that signatories undertake "to give their help in the identification, protection, conservation and preservation of the cultural and natural heritage [sites on the World Heritage List or the List of World Heritage in Danger] if the States on whose territory it is situated so request,"[164] and "not to take any deliberate measures which might damage directly or indirectly the cultural and natural heritage [sites on the World Heritage List] situated on the territory of other States Parties to this Convention."[165] Presumably, the omission of the at-risk sites on the List of World Heritage in Danger (discussed later in the chapter) from the last clause was not intended to condone the deliberate damage of those sites because all of those sites would necessarily be on the primary World Heritage List as well.

The WHC includes the well-intentioned but controversial concept of transitional zoning, or "buffer zones." The idea is that listed World Heritage sites should be surrounded by concentric regions of graduated restrictiveness to provide a margin of safety around the sites themselves. Whenever necessary for proper conservation, "an adequate 'buffer zone' around a property should be provided and should be afforded the necessary protection. A buffer zone can be defined as an area surrounding the property which has restrictions placed on its use to give an added layer of protection."[166] Of course, by expanding the territory subject to increased regulation beyond the actual formal boundaries of a listed site (such as a national park, wildlife refuge, or wilderness area), the buffer zone principle can be seen as an encroachment on the private property rights of individual landowners. This then contributes to the disputatious nature of many WHC listing proposals, as citizens fight to defend their property interests from indirect erosion.[167]

The application for a site to be inscribed on the World Heritage List must come from the country itself.[168] Moreover, no site may be placed on the list without the consent of the nation concerned.[169] An application for listing must also include a plan detailing how the site is already managed and protected in national legislation, including a demonstration of "full commitment" as evidenced by legislation, staffing, and plans for management and funding.[170] There is also a requirement that all nonfederal owners of the site concur in the nomination for listing. The World Heritage Committee[171] meets once a year and examines the applications on the basis of technical evaluations. These independent evaluations of proposed cultural and natural sites are provided by two advisory bodies, the International Council on Monuments and Sites (ICOMOS) and the World Conservation Union (IUCN), respectively.[172]

As with the Law of the Sea Treaty, the lack of any mechanism to inscribe sites that are *beyond* the territorial limits of any nation is a very serious defect when it comes to marine hotspots. Although it is possible to list marine sites in the coastal areas, such as near-shore coral reefs, the WHC cannot touch vital pockets of biodiversity in the high seas. This is ironic, given that such remote oceanic sites are perhaps the archetypal example of treasures that are truly of "outstanding universal value" and the common heritage of all of humankind, and not merely the property of any individual nation. But because these areas "belong" to no specific nation, there is no one under the WHC with the legal authority to claim them for the common benefit of humankind. This is a sad and devastating defect.

But what about those areas, including marine hotspots, that do providentially fall within the territories wherein they are eligible for WHC protection? Would the WHC afford a meaningful level of protection? The World Heritage List has grown to a formidable size. As of August 2005, the list included 812 sites of "outstanding universal value" in 137 nations.[173] Of these 812 sites, 628 are denominated as "cultural," 160 as "natural," and 24 as "mixed."[174] Two of the eight new natural sites added to the World Heritage List in July 2005 are the Coiba National Park and its Special Zone of Marine Protection off the coast of Panama and the Gulf of California. This is a positive development illustrative of the potential for the WHC to assist in marine hotspot identification and preservation.[175] A glance at the map of World Heritage sites quickly reveals that there a number of other locations that are either islands, coral reefs, or other areas of significance to marine biodiversity.[176]

The World Heritage List includes sites that fall within the terrestrial and marine hotspots, albeit sites that usually amount to only a small fraction of the territory that each hotspot actually embraces on the basis of the scientific evidence alone. Notably, given the prominent representation of tropical forests in the hotspots, the list features more than three dozen separate tropical forest sites, which in the aggregate encompass over 30 million hectares of territory. Of these sites, at least twenty-three are national parks within their respective nations, and over a dozen more are reserves or sanctuaries of one type or another. In this way, the WHC has often functioned to lend some degree of additional support to areas that had previously been identified and set apart by the host nation as an important natural property. But clearly there is far more potential than actual focus on marine hotspots to date under the WHC.

There is a World Heritage Fund established under Article 15 that provides limited financial support to nations in furtherance of the WHC's purposes. The fund, which is set up as a trust fund, is to receive compulsory and voluntary contributions from the WHC signatories, as well as from several other sources.[177] Specifically, Article 15.3 provides, in pertinent part:

> The resources of the Fund shall consist of: (a) compulsory and voluntary contributions made by States Parties to this Convention; (b) contributions, gifts or bequests which may be made by: (i) other States; (ii) the United Nations Educational, Scientific and Cultural Organization, other organizations of the United Nations system, particularly the United Nations Development Programme or other intergovernmental organizations; (iii) public or private bodies or individuals; (c) any interest due on the resources of the Fund; (d) funds

raised by collections and receipts from events organized for the benefit of the fund; and (e) all other resources authorized by the Fund's regulations, as drawn up by the World Heritage Committee.[178]

This enables the World Heritage Fund to receive contributions from a wide range of donors, including private individuals, nongovernmental organizations, and any nation. The WHC also directs states parties to "consider or encourage the establishment of national public and private foundations or associations whose purpose is to invite donations for the protection of the cultural and natural heritage"[179] as defined in the WHC. The overarching concept is to broaden the scope of possible funding sources; it also empowers the WHC to employ innovative and unconventional ideas to augment the funds available for preservation of the natural and cultural resources it seeks to safeguard. Although at present this is still largely untapped potential, the potential is nonetheless spelled out in the WHC, which sets the foundation for future progress.

The World Heritage Committee determines the acceptable uses for the fund's resources and "may accept contributions to be used only for a certain programme or project, provided that the Committee shall have decided on the implementation" of such an initiative.[180] No political conditions may be attached to contributions made to the fund.[181] In other words, interested individuals and groups, including nongovernmental organizations (NGOs), have some ability to target their donations to certain favored projects, such as the preservation of a particular sector of a hotspot. This could be a useful tool for harnessing the power and money of activists, philanthropists, and public interest groups in the WHC's efforts to assist certain sites on the World Heritage List.

With regard to the signatories to the WHC, the amount of "compulsory" contributions to the fund is discussed in Article 16, paragraph 1:

Without prejudice to any supplementary voluntary contribution, the States Parties to this Convention undertake to pay regularly, every two years, to the World Heritage Fund, contributions, the amount of which, in the form of a uniform percentage applicable to all

States, shall be determined by the General Assembly of States
Parties to the Convention, meeting during the sessions of the
General Conference of the United Nations Educational, Scientific
and Cultural Organization. This decision of the General Assembly
requires the majority of the States Parties present and voting, which
have not made the declaration referred to in paragraph 2 of this
Article. In no case shall the compulsory contribution of States
Parties to the Convention exceed 1 percent of the contribution to the
regular budget of the United Nations Educational, Scientific and
Cultural Organization.[182]

However, Article 16, paragraph 2, allows parties to issue a
"declaration" that they will not be bound to contribute to the World
Heritage Fund in the manner provided by paragraph 1. The United
States is one of the nations that has exercised the option to excuse
itself from contributing to the World Heritage Fund under Article
16.1. Again, I only wish that our income tax laws contained a similar
do-it-yourself exception. I may be going out on a limb here, but I
believe there would be more than a few such "declarations" filed
within a microsecond of the creation of this option.

Strangely, paragraph 4 of this same article of the WHC directs
that contributions from parties that have made this declaration "shall
be paid on a regular basis, at least every two years, and should not be
less than the contributions which they should have paid if they had
been bound by the provisions of paragraph 1 of this Article."[183] In
any event, sanctions for nonpayment of either "voluntary" or
"compulsory" contributions are quite limited:

Any State Party to the Convention which is in arrears with the
payment of its compulsory or voluntary contribution for the current
year and the calendar year immediately preceding it shall not be
eligible as a Member of the World Heritage Committee.[184]

Requests for international assistance for the preservation of WHC
properties are made under Article 19, and the funds are to be granted
only for duly listed sites, pursuant to Article 20. There is also technical
assistance and training available,[185] which, if offered in conjunction

with sufficient levels of financial aid, might be instrumental in effecting meaningful protection for World Heritage Sites. Article 22 specifies that assistance to sites on the World Heritage List may take the form of: studies concerning the artistic, scientific, and technical problems raised by the protection, conservation, presentation, and rehabilitation of the site; provision of experts, technicians, and skilled labor to ensure that the approved work is correctly carried out; training of staff and specialists at all levels in the field of identification, protection, conservation, presentation, and rehabilitation of the site; supply of equipment which the nation concerned does not possess or is not in a position to acquire; low-interest or interest-free loans which might be repayable on a long-term basis; and the granting, "in exceptional cases and for special reasons, of non-repayable subsidies."[186]

Could the quantum of assistance provided under the WHC suffice to make an outcome-determinative difference for any site, including a marine hotspot? The language of the convention is characteristically vague:

> International assistance on a large scale shall be preceded by detailed scientific, economic and technical studies. These studies shall draw upon the most advanced techniques for the protection, conservation, presentation and rehabilitation of the natural and cultural heritage and shall be consistent with the objectives of this Convention. The studies shall also seek means of making rational use of the resources available in the State concerned.[187]

The text does not make any attempt to define the key terms "large scale," "detailed" studies, and "most advanced techniques." The imprecision of the standards leaves important decisions on the appropriate degree of help to the discretion of the World Heritage Committee. Similarly, the restriction in Article 25 to the effect that "only part of the cost of work necessary shall be borne by the international community" and the nation benefiting from international assistance shall contribute "a substantial share of the resources" devoted to each program or project, is not a firm, objective standard.[188] Moreover, any limitation on aid or mandate for host nation contribution implicit in Article 25 is overcome by its concluding escape

hatch, "unless [the host nation's] resources do not permit this."[189]
Very often, of course, the host nations for hotspots are in desperate
economic straits, which is a primary reason why their natural resources are imperiled in the first place. Pressures to develop and exploit nature are most acute when there are few, if any, alternatives for
a nation and its people who are struggling to maintain a bare sub-
sistence level of income.

In prescient anticipation of a shortfall of available rescue resources
and a surplus of pressing and competing needs, the WHC reflects an
attempt to set forth a system for setting priorities:

> The Committee shall determine an order of priorities for its op-
> erations. It shall in so doing bear in mind the respective im-
> portance for the world cultural and natural heritage of the property
> requiring protection, the need to give international assistance to the
> property most representative of a natural environment or of the
> genius and the history of the peoples of the world, the urgency of
> the work to be done, the resources available to the States on whose
> territory the threatened property is situated and in particular the
> extent to which they are able to safeguard such property by their
> own means.[190]

A key feature of the WHC in terms of hotspots preservation is the
set of the measures it prescribes when sites are imperiled. The World
Heritage Committee is supposed to be alerted—by individuals, nongovernmental organizations, or other groups—to possible dangers to a
site. If the alert is justified, and the problem serious enough, the site
will be placed on the List of World Heritage in Danger, which is
provided for by Article 11.4 of the WHC.[191] The List of World
Heritage in Danger is reserved for those sites already inscribed on the
primary World Heritage List "for the conservation of which major
operations are necessary and for which assistance has been requested"
under the WHC.[192] The list is to contain an estimate of the costs of
any such operations. Furthermore,

> The list may include only such property forming part of the cultural
> and natural heritage as is threatened by serious and specific dangers,

such as the threat of disappearance caused by accelerated deterioration, large-scale public or private projects or rapid urban or tourist development projects; destruction caused by changes in the use or ownership of the land; major alterations due to unknown causes; abandonment for any reason whatsoever; the outbreak or the threat of an armed conflict; calamities and cataclysms; serious fires, Earthquakes, landslides; volcanic eruptions; changes in water level, floods and tidal waves.[193]

This List of World Heritage in Danger, consisting of imperiled cultural and natural resources, is designed to call the world's attention to natural or humanmade conditions that threaten the characteristics for which the site was originally included in the main World Heritage List.[194] In theory, inclusion on the "Danger" list increases the likelihood that funds will be deemed available within the priority-setting triage scheme of Article 13.4 to make a difference in the survival of the resources in question.

The List of World Heritage in Danger included only thirty-three sites as of August 2005.[195] Many of the sites on this list are cultural/ historical resources rather than natural resources, but the list is open to both categories. The United States currently has only one site inscribed on the list—the Everglades National Park (Yellowstone National Park was also on the list for a time).[196] Several terrestrial parks and nature preserves in other nations are on the list, including the Srebarna Nature Preserve in Bulgaria; the Manovo-Gounda St. Floris National Park in the Central African Republic; the Mount Nimba Nature Reserve in the Ivory Coast/Guinea; the Virunga, Garamba, Kahuzi-Biega, and Salonga National Parks and Okapi Wildlife Reserve, all in the Democratic Republic of the Congo; the Sangay National Park in Ecuador; the Rio Platano Biosphere Reserve in Honduras; the Manas Wildlife Sanctuary in India; the Air and Tenere Natural Reserves in Niger; the Djoudj National Bird Sanctuary in Senegal; the Ichkeul National Park in Tunisia; and the Rwenzori Mountains National Park in Uganda.

The marine hotspots should be extensively represented on the List of World Heritage in Danger, on the basis of the confluence of core criteria for inclusion. If there were broader recognition and comprehension of the hotspots concept worldwide, their representation on the List of World Heritage in Danger would be far more extensive than it is now. By definition, the marine hotspots are both supremely vital repositories of much of the Earth's biodiversity, and are drastically under attack from a variety of destructive developmental forces. If anything belongs on the List of World Heritage in Danger, marine hotspots certainly do.

Unfortunately, the act of inscribing a site on either the World Heritage List[197] or the List of World Heritage in Danger can be very controversial. When Yellowstone National Park was placed on the List of World Heritage in Danger in 1995, much political furor arose. Claims were made that U.S. sovereignty had been impinged merely because the WHC had influenced President Clinton's decision to issue executive orders providing buffer zones around the park and enhancing its protection against a nearby mining operation.[198] Today Yellowstone is no longer on the list, which could be viewed as evidence that either conditions there dramatically and swiftly improved, or that political pressure forced the removal. What is your guess? A cynic might be forgiven for opining that this is corroboration of the maxim "No good deed goes unpunished."

One additional feature of the WHC could be useful under the right circumstances, albeit indirectly. Article 27 focuses on educational and informational initiatives to inform the citizenry as to the importance and fragility of World Heritage sites:

> 1. The States Parties to this Convention shall endeavor by all appropriate means, and in particular by educational and information programmes, to strengthen appreciation and respect by their peoples of the cultural and natural heritage defined in Articles 1 and 2 of the Convention. 2. They shall undertake to keep the public broadly informed of the dangers threatening this heritage and of the activities carried on in pursuance of this Convention.[199]

The evident intent is to educate the people, at all levels, within the nations that are home to the various World Heritage sites. The drafters of the WHC recognized the importance, indeed the indispensable nature, of widespread public knowledge and support of conservation efforts, particularly with regard to key natural and cultural treasures. If the people "on the ground" in these nations do not know the value of the sites with which they may interact, and are uninformed as to the dangers threatening the continued existence of the sites, they cannot be expected to personally hold them in high esteem. They cannot be expected to refrain from exploiting and damaging the sites when it is their financial self-interest to do so, let alone voluntarily devote their own time, effort, and money to the preservation of the sites. And absent this type of grassroots commitment of the citizenry, there is very little real protection that can be imposed on sites from the top down. Thus, the spirit of Article 27 is in tune with a very real and persistent problem that has plagued conservation globally, and, at a minimum, it reflects an attempt to ameliorate the situation by using understanding and information as the best antidotes to apathy and antipathy.

Unfortunately, the WHC lacks any true enforcement mechanisms. This has vitiated many of the potentially useful provisions in the convention. If a signatory fails to fulfill its obligations under the convention, it risks having its sites deleted from the World Heritage List, but this is not a sufficient deterrent for a nation that fails to demonstrate the requisite level of commitment to the principles of the WHC. Despite its terms that purport to obligate parties to refrain from undertaking acts that might directly or indirectly damage a designated resource, the WHC does not address whether sanctions may be taken against countries that violate its terms and conditions.[200] Also, while signatories are required to submit reports regarding domestic measures taken in furtherance of WHC aims,[201] there is no provision whereby a party can be penalized or sanctioned for failing to provide requested information or for submitting inaccurate or false information. As a result, reports have been less than satisfactory in many cases.[202] The WHC does not provide a dispute settlement process either.[203]

Philosophically, the WHC is quite compatible with the concept of marine hotspots preservation and may provide some assistance toward this aim, as it has in other areas.[204] Among the criteria for consideration as a "natural heritage" site is that an area be of "outstanding universal value from the point of view of science or conservation."[205]

This definition is tailor-made for hotspots. And, as we have seen, the factors that determine eligibility for inclusion in the top-priority subsidiary list, World Heritage in Danger, are also entirely consonant with the very definition of a hotspot.

However, this philosophical fit is spoiled by the lack of meaningful "teeth" to enforce its provisions; loss of WHC listing of a nation's resources is the only sanction for noncompliance.[206] This is akin to punishing someone who beats his pet dog by telling him his dog will no longer be allowed to have a license. Moreover, the WHC leaves it up to individual nations to recommend their own resources for inclusion in the World Heritage List and prohibits inclusion without the consent of the host nation. A nation that is disinclined to preserve its hotspot would be unlikely to nominate it for the list, and would probably veto any attempt by outsiders to inscribe it. After all, is it really true that there are only thirty-five places (whether cultural or natural) in the *entire world* that properly qualify for the List of World Heritage in Danger? If not—if there are many more that deserve that designation—then there must be powerful disincentives and structural defects at work that have artificially depressed the number of treasures thus inscribed.

One crucial problem is that there is no mechanism for listing sites outside of national jurisdiction. There also seems to be an open question on the limits of a nation's ability to list marine areas. What, exactly, constitutes a nation's jurisdiction: within the territorial seas, or out to the limits of the EEZ or continental shelf? A coastal nation does not even possess complete sovereignty over the relatively near-shore areas—the nation possesses only sovereign rights. In 1972, when the WHC was adopted, UNCLOS and its jurisdictional regime were still twelve years from adoption; thus, the WHC does not specifically address these key marine areas. Regarding areas clearly outside of national jurisdiction, it seems axiomatic that such global common areas should be studied, protected, and funded by the global community. Yet the WHC does not address these areas at all. As with UNCLOS, this is a missed opportunity of tragic proportions.

Coupled with the low level of financial assistance currently available for preservation efforts, these core features of the WHC have rendered it, in its present form, ineffective in protecting the hotspots, whether in the marine or the terrestrial sphere, but most emphatically in the marine realm. However, the potential is there for the WHC to make a meaningful contribution someday if significant global attention is focused on the undeveloped potential of the WHC philosophy and the requisite amendments are made in the future.[207] Until then, it is most unfortunate that some of the crown jewels of Earth—the marine hotspots—lie unprotected under the sonar of the WHC, and cannot even be inscribed on the List of World Heritage in Danger.

LONDON DUMPING CONVENTION

I will now consider the Convention on the Prevention of Marine Pollution by Dumping of Waste and Other Matters (London Convention, or London Dumping Convention), which was designed to provide the basic framework for global control of the deliberate disposal of all wastes into the oceans.[208] This was deemed necessary because of the widespread practice of collecting wastes that had initially been generated on land, loading them onto a ship or barge, and then taking them out to sea for the express purpose of dumping— essentially treating the oceans as a giant toilet/garbage disposal combination. The convention also includes deliberate disposal from aircraft, platforms, and other humanmade structures within its prohibitions, to the same extent as dumping from vessels.

The convention bans the intentional disposal of certain hazardous substances,[209] and requires a permit from the coastal nation for the dumping of other substances.[210] Amendments to the original convention banned the dumping of nuclear waste[211] and regulated the incineration of waste at sea.[212] Each state party has the duty to enforce the convention within its jurisdiction.[213] However, responsibility for enforcement on the high seas lies with the nation where the dumping vessel is registered.[214]

In actual practice, bureaucratic control over the convention lies with the International Maritime Organization (IMO) in London. The IMO is an agency within the United Nations, and it serves as the secretariat for the convention. It is charged with serving as a central repository for dumping permits issued by all the governments of signatories. The IMO also disseminates information to the convention's signatories and holds periodic consultative and scientific group meetings.[215]

The convention's primary tool is a permit system. Irrespective of whether a given waste is actually intended to be dumped within a signatory's territorial waters, the state is obligated to have a permit for all ships and barges that load in its ports or its waters for purposes of dumping. This permit requirement applies to a signatory's own vessels, as well as those of other nations, when the signatory plans to dump wastes in any of the world's oceans, wherever situated.

Annex I of the convention is a list of the substances that are banned from ocean dumping because of their considerable potential to harm the marine environment. This is the "blacklist."[216] Even for blacklisted substances, however, the ban is not absolute. There can be exceptions, allowing ocean dumping of otherwise blacklisted substances, in instances of "emergencies, posing unacceptable risks relating to human health and admitting no other feasible solution." Also, the blacklist does not ban dumping of wastes that are "rapidly rendered harmless by physical, chemical, or biological processes in the seas." This is a major loophole big enough to steer a cruise ship through.[217] Plus, when blacklisted substances are found only as "trace contaminants" in wastes such as sewage sludge or dredged spoils, the ban does not apply, even though sufficiently prolonged accretion of "trace" amounts of heavy metals in a particular place has the potential to add up to a considerable threat over time.

The Annex II counterpart to the blacklist is the "graylist." The graylist consists of wastes containing "significant amounts" of somewhat less harmful materials. Graylisted substances, while still of concern, are thought to be less harmful to the marine environment than those on the blacklist, and they may be dumped in the ocean so long as "special care" is taken as to site selection, packaging of the wastes, monitoring, and choice of disposal methods to mitigate harmful impacts.[218]

Those substances that are not listed in either Annex I or Annex II are governed by Annex III, and may be dumped in compliance with a general permit. An Annex III permit issued by any signatory is supposed to reflect careful consideration of specified environmental protection criteria. Such criteria include possible effects on marine biodiversity, as well as effects on other uses of the oceans, and are to reflect the characteristics and composition of the particular waste in question. Of possible relevance to marine hotspots is the requirement that a permit also address a description of the intended disposal site and the practical availability of land-based disposal alternatives.

A more stringent regulatory scheme is found in the 1996 Protocol to the London Convention, which parties to the original convention may join if they wish. This protocol is not yet in force because it has not garnered the requisite ratification by twenty-six states, including at least fifteen Parties to the London Convention; as of this writing, the protocol has been ratified by twenty-one states.[219] For the nations that join the protocol, it supersedes the 1972 Convention. The protocol adopts the "precautionary approach" and "polluter pays" concepts, and allows ocean dumping only for certain listed wastes, and only under the terms of a permit.

Reflecting these principles, the 1996 Protocol embodies a major structural revision of the convention in what has become known as the "reverse list" approach. Under this rubric, rather than prohibiting the dumping of certain specifically listed hazardous materials, the parties are obligated to prohibit the dumping of any waste or other matter that is *not* listed in Annex 1 ("the reverse list") of the 1996 Protocol.[220] In other words, the presumption is that ocean dumping is banned unless there is a valid permit, rather than that all dumping is allowed except for certain special cases. This shift in the default position marks a huge about-face in approach. Dumping of wastes or other substances on this reverse list requires a permit. Parties to the protocol are further obliged to adopt measures to ensure that the issuance of permits and permit conditions for the dumping of reverse list substances complies with Annex 2 (the Waste Assessment Annex) of the protocol. The substances on the reverse list include dredged material; sewage sludge; industrial fish-processing waste; vessels, offshore platforms, or other humanmade structures at sea; inert, inorganic geological material; organic material of natural origin; and bulky items including iron, steel, concrete, and similar materials for which the concern is physical impact. Dumping is limited to those circumstances where such wastes are generated at locations with no land-disposal alternatives. The protocol contains an outright prohibition of incineration of wastes at sea (except for emergencies), bans the export of wastes to other nations for purposes of ocean dumping or incineration, and establishes some dispute resolution and technical cooperation/assistance procedures.

Ocean dumping constitutes approximately 10 percent of the pollution to the ocean.[221] While worldwide these effects may not be extensive, localized effects are most likely pronounced. The majority of substances dumped at sea fall within the categories of sewage sludge and dredge spoils.[222] Even if these materials are "clean," the impacts in a particular locale can be devastating. Significantly, most dumping takes place close to shore for reasons of economy and convenience, and that exacerbates the risk to coral reefs and other centers of biodiversity near land. Disposal is concentrated in these relatively small areas rather than evenly distributed over the ocean's vast expanses, and this concentration of waste coincides with concentrations of biodiversity in many continental-shelf habitats. Benthic communities, hydrothermal vents, or other undiscovered communities could also be smothered by this refuse. Similarly, in the case of sewage sludge, the resulting nutrient enrichment could cause short-term productivity gains followed by a die-off and a resulting low-oxygen environment.[223]

While incorporating some very progressive features, the convention contains major weaknesses and loopholes. The convention allows for substances to be added to the list of banned substances with a two-thirds majority vote, but an opt-out clause allows states to avoid being legally bound to provisions to which they do not wish to adhere.[224] Reporting and enforcement activities are left largely to the signatory, and many nations do not take these requirements very seriously.[225] For example, Russia continued to dump nuclear waste at sea after the convention banned all dumping of nuclear waste.[226]

Additionally, like all treaties, only nations that have signed on, and in this particular case, nations that have not opted out of specific provisions, are bound by the London Dumping Convention. Many developing nations have not signed or ratified the treaty, and the convention provides little incentive to spur these nations to join.[227] Finally, this agreement is a prime example of the single-use focused treaty. While possibly effective in limiting dumping, it only focuses on one threat to unique systems, without addressing real protection in any meaningful or comprehensive way.

CONVENTION ON INTERNATIONAL TRADE IN ENDANGERED SPECIES

The Convention on International Trade in Endangered Species of Wild Flora and Fauna (CITES),[228] as the name implies, deals with international trafficking in endangered species. CITES entered into force on July 1, 1975, and as of August 2005 had attracted 169 parties.[229] These nations act by banning commercial international trade in an agreed list of endangered species and by regulating and monitoring trade in other species that might become endangered.[230]

CITES is essentially an international version of the United States' Endangered Species Act (ESA)[231] provisions that prohibit such trafficking. In fact, the ESA is the means by which the United States fulfills its obligation to implement CITES. Under CITES, export and import of endangered species requires a government permit which can be granted only if the following conditions will be met: trade will not be detrimental to survival of the species, the specimen was not obtained contrary to applicable nature protection laws, and shipment will not result in injury or cruel treatment.[232] Appendices set forth categories of endangered species, with the most vulnerable being most severely regulated.[233] In implementing CITES, the European Community sought to achieve uniform protection within the Community and, for some sensitive species, provided even stricter protection than the convention required.[234]

CITES provides some enforcement mechanisms, such as the Article VIII requirement that parties take "appropriate measures" to enforce CITES provisions, including assessing penalties on violators, confiscating illegal trade, and imposing fines for the costs incurred from the confiscation of illegal trade.[235] Article VIII also requires parties to submit implementation reports to the CITES secretariat annually.[236] Additionally, Article XIII allows the secretariat to bring noncompliance matters to the attention of the parties involved when the secretariat is convinced that treaty provisions have not been "effectively implemented."[237] There is a dispute resolution procedure as well.[238] However, these enforcement tools have been criticized on multiple grounds as falling "far short of establishing a coherent, uniform system for interpreting and enforcing CITES."[239] One problem is that use of the word "recommendations" in Article XI indicates that the enforcement mechanisms are not legally binding.

Even on its own terms as a species-specific treaty, CITES has garnered decidedly mixed reviews.[240] In part, this stems from the fact that CITES neither prevents species from harm, nor does it protect them from complete domestic elimination within any given nation; CITES only regulates international events. Significantly, CITES allows parties to take formal legal reservations as to any species listed in Appendices I–III or any parts/derivatives specified therein, either at the time the nation becomes a party or upon amendment to an appendix.[241] Such reservations allow reserving parties to be treated as nonparties with regard to trade in the applicable species or its parts/derivatives, unfettered by CITES requirements. Reservations have been used frequently under CITES, to the detriment of listed species.[242]

In terms of marine hotspots protection, the CITES might only be useful in preventing international trade stemming from poachers and those who profit from poaching. This, of course, does not directly affect the marine hotspots as such, nor does it provide overarching protection for the ecosystems or habitats in which endangered species live, although it certainly is a worthy and important provision in its own right. Similar to the ESA, CITES operates one species at a time, and cannot help until a species approaches its deathbed—a decidedly ineffective and reactive approach to aiding even individual species, let alone entire ecosystems or hotspots.[243] Rather than a holistic focus on the entire ecosystem mosaic, CITES aims at reinforcing loose tiles, one by one. It is worthwhile, but hardly the big-picture proactive curative the hotspots demand.

The focus on individual species also places CITES in the same category as the Whaling Convention, or the Bonn Convention on migratory species, in that it is limited, by its own terms, to a specific subset of the total biodiversity of the world. Any agreement that aims only at individually listed species, or whales, or migratory marine species, cannot be relied on to hit the broader target of marine biodiversity hotspots; that is not the intended effect of such treaties, and it is unrealistic to expect it of them. Thus, CITES—as with the other conventions mentioned—is best viewed as a gap-filling supplement to other legal measures directed at marine hotspots preservation. CITES is not the answer in and of itself, nor was it intended to be.[244]

And so, after looking at all of these international laws, and after pondering the significance of these many words, I think you can begin to see why I made the blunt and cynical statements at the beginning of this chapter. If we want public posturing, impressive displays of action-free concern, and loads of make-believe, then the international legal solution to our mass extinction will do just fine, thank you very much. But the fact is that all these treaties and agreements have been around long enough that we should be seeing some good-news results by now, if they truly were a realistic response to this extinction spasm. Nothing could be further from the truth. All indications, as discussed in excruciating detail in chapter one, are that the situation is only growing worse every year, while all these stacks of laws stand by silently and motionlessly. And that is all they do, because that is all they *can* do.

The tragic flaw of international law is that it only has meaning for those who would do the right thing anyway, and has no power to force real change upon those inclined to disregard it. For the nations that want to protect marine life, no international law is necessary. For the nations that want the opposite, no international law will suffice.

THREE

Finding or Losing Nemo, One Nation at a Time

In this chapter I will shift focus from international laws to the individual laws of some of the nations with notable marine biodiversity resources—including marine hotspots—within their jurisdiction. Given the weaknesses, loopholes, gaps, and other flaws in the international-law safety net, we need to establish whether the key nations have stepped effectively into the breach with laws of their own. Sadly, although there is no lack of national laws pointed in some measure at preserving marine biodiversity, the total is generally less than the sum of its parts for most nations.

Rather than exhume the remains of failed efforts from dozens of nations, I will focus primarily on two relatively encouraging success stories, the United States and Australia. There certainly is room for improvement even with these countries, but at least they have led the way and made a decent effort toward legal protection for marine biodiversity. If they and other nations would build on this foundation, along the lines I will outline in chapter four, there could be real progress.

One factor that has undermined the quality of legislation in virtually every country is the tendency for legislatures to do nothing until the solid organic waste matter hits the oscillating air circulation apparatus. As Isaac Newton pointed out centuries ago (in a different context), objects at rest tend to remain at rest. Legislative inertia is a powerful force on the side of doing nothing, at least until and unless there is some perceived crisis along the lines of all hell breaking loose; even Congress can understand that, of all the things one does not want to break loose, "all hell" heads the list. But short of a headline-hogging, outrage-spawning emergency, the lawmakers will often not be particularly inclined to devote their attention to topics that, like marine biodiversity, stay well below the public's sonar. Therefore, most of the laws on the books were either passed in haste in the immediate aftershocks of a well-publicized dust-up, or were primarily a response to some issue other than life in the oceans.

UNITED STATES

United States domestic legislation regarding marine biodiversity has followed a piece-by-piece approach, focusing one at a time on particular uses or activities, individual species, or individual areas. The lack of an umbrella statute with comprehensive coverage has resulted in some inefficiencies and gaps, but there is certainly no lack of applicable laws in terms of sheer volume. Quantity may not be a substitute for quality, but at least it demonstrates some level of leg-islative commitment to the cause, even if much of it falls within the "window dressing" category.

In the United States, over 140 laws pertain to the oceans and coasts, 43 of which are generally considered major.[1] There is a pleth-ora, indeed a veritable shipload, of laws relevant to protecting marine biodiversity, including habitats such as seamounts, coral reefs, and vent areas. One of the most notable statutes is the Marine Protection, Research, and Sanctuaries Act of 1972 (MPRSA), which, like many other laws, has a name more promising than its true value.[2]

The MPRSA has two distinct objectives. First, Title I of the act, which is often referred to as the Ocean Dumping Act, regulates the intentional ocean disposal of materials. Title I prohibits all ocean dumping, in any ocean waters under U.S. jurisdiction by any U.S. vessel, or by any vessel sailing from a U.S. port, except as authorized by permit.[3] No dumping of radiological, chemical, and biological warfare agents, high-level radioactive waste, or medical waste is permitted.[4] However, the administrator of the Environmental Protection Agency may issue a dumping permit (except for dredge spoils) where the administrator determines that such dumping will not unreasonably degrade or endanger human health, welfare, the marine environment, ecological systems, or economic potentialities.[5] Permits for the dumping of dredge spoils are issued by the Secretary of the Army based on the same criteria.[6]

The superficially strict prohibitions on ocean dumping are frequently swallowed up by the exceptions in actual practice. To illustrate, each year approximately 60 million cubic yards of dredged material are disposed of in the ocean at designated sites.[7] The statute provides for civil and criminal penalties, and allows individuals to bring a citizen suit against any person, including the United States, for violation of a permit or other prohibition, limitation, or criterion issued under Title I of the act.[8]

Title III of the MPRSA authorizes the establishment of marine sanctuaries within areas of U.S. national jurisdiction. This is the type of provision that, at least theoretically, is most useful for preserving marine hotspots—the marine equivalent of wilderness areas, wildlife refuges, and other forms of terrestrial protected areas. The act authorizes the secretary of commerce, acting through the National Ocean Service, to designate any discrete area of the marine environment as a national marine sanctuary and promulgate regulations implementing the designation.[9] The governing factors include "the area's natural resource and ecological qualities, including its contribution to biological productivity, maintenance of ecosystem structure, . . . and the biogeographic representation of the site."[10] The act requires notification of affected federal, state, and local governments, as well as preparation of an Environmental Assessment under NEPA, and public notice/comment, prior to designation of a marine protected area (MPA).[11]

But as alluded to in chapter two during the discussion of UNCLOS, most MPAs are not strict sanctuaries (along the lines of national wilderness areas) off-limits to virtually all human interference. Instead, they are more akin to national forests, in that they are generally managed for multiple uses—including recreation, education, commercial fishing, and shipping—not just biodiversity preservation.[12]

The National Ocean Service is required, under the regulations applicable to most MPAs, to allow public and private uses "to the extent compatible with the primary objective of resource protection."[13] Given the many governmental (as well as private and commercial) activities that therefore take place within MPAs, the MPRSA imposes a consultation requirement analogous to the provisions of Section 7 of the Endangered Species Act (ESA). Federal agencies must consult with the National Ocean Service with respect to actions they plan to undertake, whether inside or outside an MPA, that are likely to "destroy, cause the loss of, or injure any sanctuary resource."[14] This procedural requirement can result in the recommendation of "reasonable and prudent alternatives" to the proposed action, but the action agency is free to depart from such alternatives so long as it justifies this in a written statement.[15]

There are other respects in which the MPA concept is less than fully realized, to put it charitably. For example, the Magnuson-Stevens Fishery Conservation and Management Act[16] provides for eight Regional Fishery Management Councils to be established, and these councils have the power to decide whether commercial fishing regulations are needed in any given MPA and, if so, to draft them themselves. The fishery management plans developed under Magnuson-Stevens are to be designed to meet nationally set goals that balance multiple concerns (including the well-being of the fishing industry), and these goals in turn drive certain consequences as they affect commercial fishing regulations.

Under the MPRSA, these Fishery Management Council self-written regulations "shall be accepted and issued as proposed regulations by the Secretary unless the Secretary finds that the Council's action fails to fulfill the purposes and policies [of the MPRSA] and the goals and objectives of the proposed designation [of an MPA]."[17] This scheme renders it improbable that commercial fishing will be limited to a meaningful extent within or near an MPA; in fact, most litigation regarding MPA regulations does not deal with commercial fishing restrictions, but rather with limitations on personal recreational activities.[18] The government may as well post "water of many uses" signs on buoys along the perimeter of MPAs, like to those that appear in national forests.

The Office of the National Marine Sanctuaries, under the National Ocean Service, administers all U.S. MPAs, which currently number thirteen.[19] In addition to the MPRSA, legal authority comes from Executive Order 13,158, "Marine Protected Areas," which directs federal agencies to conserve key marine resources through a variety of measures related to MPAs.[20] There are a number of other statutes in addition to the MPRSA that provide authority to declare MPAs under some circumstances.[21] A Marine Protected Areas Center, under NOAA, has been created pursuant to this executive order so as to coordinate efforts to implement the order and provide support.[22]

The combined effect of the MPAs is certainly better than nothing, but the multiple-use, sustained-yield approach and the statutory concessions to commercial fishing guarantee that our MPAs are less effective than they might be.[23] Although the efficacy of marine sanctuaries is a matter of some scientific debate, it is incontrovertible that a loophole-ridden sanctuary is less protective of biodiversity than a stringent one.[24]

The Coastal Zone Management Act (CZMA)[25] is another major U.S. statute typical of nations with significant coastal resources such as coral reefs, barrier islands, and other related areas. Essentially a planning/procedural statute, CZMA has some utility in its ability to focus attention on marine resources, including centers of biodiversity. In fact, CZMA explicitly directs the establishment of an interagency task force to examine the causes and effects of harmful algal blooms and hypoxia.[26] It also directs the establishment of a National Coastal Resources, Research and Development Institute, among other things.[27]

Along these lines, CZMA serves to prod coastal states (including those bordering an ocean or the Gulf of Mexico) within the United States to develop and coordinate programs for management of their coastal zone resources, with provision for federal funds and technical assistance.[28] Federal activities are to be "consistent" with these state management programs.[29]

CZMA relies to a great extent on the states to implement federal policy through state-designed land-management decision-making mechanisms. This is one of the most extensive regimes of federalism in all of environmental law, with a very large role for the states to play. In fact, all that CZMA asks of the states is to take steps to address the national goals satisfactorily, and, if this is done, the states have wide latitude in how they craft their land-use and water-use management frameworks. The act contains built-in inducements to encourage the states along these lines; it makes federal grant money available for the implementation of any approved coastal management program. Theoretically, this provides plenty of room for individual states to experiment with whatever legal arrangements seem best fitted to their particular circumstances and concerns, although it also makes it more difficult for the federal government to impose much in the way of uniform national standards.

Additionally, through its "consistency" determinations, CZMA requires federal agencies to conduct their programs and activities in harmony with the state-developed management programs, "to the maximum extent practicable."[30] Now, where have we seen that phrase before? Oh yeah, in all those international laws discussed in chapter two. As a result of this built-in wiggle room, the ecosystem management system of CZMA is decentralized to a high degree, with considerable variation from state to state.[31] Some states have plans that are more stringent than the national program envisioned under CZMA, while others adhere only to a minimal degree to the national goals.[32] And within any state, a particular federal project can be approved that might not be approved in other coastal states; the converse is also true.

As with NEPA and other procedural/planning statutes, CZMA lacks the power to dictate substantive changes in any direct manner. It is always helpful to foster research and education, and to require that actions be consistent (albeit only—you guessed it—"to the maximum extent practicable," a standard which obviously leaves an escape hatch hanging wide open for any given project) with an adequate management plan, but these requirements are properly only one small part of the answer to the plight of marine biodiversity, even within the coastal zone of the United States.

The Endangered Species Act does have substantive teeth, unlike CZMA, but it, too, has features that greatly limit its utility in safeguarding marine hotspots. The ESA focuses on protecting individual species, including marine species, that have been individually listed as threatened or endangered.[33] The Secretary of Commerce is responsible for officially listing marine species to be protected. The ESA contains strict provisions making it unlawful for anyone to "take," meaning "harass, harm, pursue, hunt, shoot, wound, kill, capture or collect, or to attempt to engage in any such conduct."[34] Additionally, the ESA contains provisions for protecting "critical habitat" necessary for each listed species, and requires federal agencies whose actions may "jeopardize" a listed species to consult with the appropriate agency[35] regarding alternatives.[36]

The ESA has numerous deficiencies in terms of protecting marine biodiversity.[37] First, species are individually listed, requiring that they be identified first. As described in chapter one, there are probably millions of unidentified species in the oceans, many of which are severely threatened, yet they cannot be listed and protected by the ESA because no one knows they exist. Second, the ESA fails to take an ecosystem-protection approach in favor of individually listed species. The act provides no positive inducements that might incentivize private individuals to safeguard imperiled species; rather, it can severely limit the freedom of property owners regarding the use of their own property, without any compensation. And finally, the provisions of the ESA do not go into effect until a species is already severely threatened, thus limiting the species' opportunity to recover.[38] By holding back all of its protections until a species is already on its deathbed, the ESA virtually guarantees that its aid will be ineffective. Many of the species that crawl onto its lists are already the "living dead" and it is just a matter of time until they finally become extinct, no matter how stringent the eleventh-hour safeguards might be.

This last point deserves a little more attention. The ESA, like its international relative CITES, is by its own terms a *reactive* law rather than a *proactive* one. Because it cannot be triggered until a species is in rather desperate condition, it can at best be used as the legal equivalent of an emergency room, and at worst hospice care, perpetually in crisis mode. It reflects the unfortunate human tendency to procrastinate, to delay taking action until a calamity erupts. Nowhere is this more counterproductive than in the area of biodiversity preservation. Because Congress does not understand the concept of species becoming so decimated that they are at some point "committed to extinction," beyond any possibility of full recovery, the ESA was inadvertently written to prevent anything more than a postponement of the inevitable. It is like a tragic ancient Greek myth where the gods decide they want to save some living things, but the method they select cannot possibly do anything until it is too late.

Congress often remains mired in inertia until and unless some catastrophe blows up in its face and screams for attention. "Inaction until emergency" might conceivably be a passable default option for other areas of legislative concern, but it is decidedly the worst way to go about conserving biodiversity. Once a species' population sinks below a certain level, it is doomed to extinction, to an absolute certainty—condemned with no chance of pardon—although the ultimate demise of the last remaining stragglers might still be thousands of years off. And Congress drafted the ESA to insist that a species be so devastated that it is already "threatened" or "endangered" as a prerequisite for qualifying for the act's protections. If Congress had affirmatively intended to enact a law that would *only* provide hospice care to the living dead, it could not have done a better job.

The Marine Mammal Protection Act of 1972[39] is a close relative of the ESA, with some of the same claws and flaws. Like the ESA, the MMPA has a prohibition on "takings" and importation of marine mammals, and implementation is divided between Fish and Wildlife Service and National Marine Fisheries Service. But unlike the ESA, there is no listing requirement. The MMPA applies to all "marine mammals," be they whales, dolphins, seals, manatees, etc., regardless of degree of extinction threat, but instead focuses on achievement of certain population levels.[40] The act establishes a Marine Mammal Commission, an independent advisory board that is responsible for reviewing and studying U.S. activities. The Marine Mammal Commission conducts a continuing review of the stocks of marine mammals, for their protection and conservation.[41]

The "primary objective" of the MMPA is officially "to maintain the health and stability of the marine ecosystem."[42] Although certainly a laudable goal, it is light years beyond the power of the MMPA to achieve. The MMPA, by its own terms, is limited to marine mammals, and maintaining an "optimum sustainable population" (OSP) thereof. The definition of OSP in the act is "the number of animals which will result in the maximum productivity of the population or species, keeping in mind the optimum carrying capacity of the habitat and the health of the ecosystem of which they form a constituent element."[43] Because OSP aims at populations, and not strictly habitat protection, it is a very poor method of protecting marine hotspots. If the ESA is inadequate for hotspots preservation because of its focus on individual imperiled species, the MMPA is even more inadequate; at best, it is only a small part of the solution to the hotspots' crisis because it lacks the concept of "critical habitat" as featured by the ESA.

Instead of any habitat-based protections, the MMPA relies on, as its primary management tool for attaining OSP, a complete moratorium on "the taking and importation of marine mammals and marine mammal products."[44] Worse, even this so-called "complete cessation" of takings is incomplete. The prohibitions on takings and importations can be waived under several circumstances, mostly hinging on whether the population of a given species is "depleted" (i.e., "a species or population stock...below its optimum sustainable population," or "listed as an endangered species or a threatened species under the Endangered Species Act"[45]). And even for depleted species, the applicable secretary has the authority to permit incidental takings "in the course of commercial fishing operations," similar to the incidental take permit provision of the ESA.[46]

Out of the hodgepodge of statutes with some connection to marine biodiversity, the United States Coral Reef Task Force was created by executive order in 1998. The National Oceanographic and Atmospheric Administration (NOAA) administers this task force, with significant involvement from EPA.[47] The task force reflects the fact that the U.S. government has recognized the lack of, and the need for, some overarching authority over key hotspots of marine endemism.[48] Certainly, even a high-level task force is no adequate substitute for comprehensive, effective legislation, but at least it represents some movement in the right direction. It is the classic governmental variation on the time-honored game of kick-the-can: When you feel you must do something, but do not know what to do, form a task force.

AUSTRALIA

Australia, along with the United States, has become a world leader in pointing the way to a responsible legal approach to marine biodiversity. Australia has assembled a formidable set of laws on point, and views the entire enterprise through the lens of an overarching, albeit still developing, oceans policy. Other nations would do well to emulate this comprehensive program.

The primary feature of Australia's marine legal regime is an ecosystem-based planning and management system within a system of Regional Marine Plans (still being formulated).[49] By centering its legal approach on the biologically sound concept of marine ecosystems, Australia has recognized that the standard, procrustean, one-size-fits-all method of governance fits the particular needs of the widely variegated marine environment about as well as Cinderella's petite glass slipper fit her big-footed stepsisters. The network of Regional Marine Plans will allow Australia to tailor its legal response to the specific needs of each key ecosystem.[50]

Australia's ambitious and innovative plan will take years to be fully developed, but there have been encouraging first steps. For example, the vital South-east Regional Marine Plan has featured some detailed impacts assessment to gain a better grasp of the condition, needs, and desirable improvements in this region.[51] This is one of six such reports that will form the foundation for Australia's Oceans Policy. A "Legal Framework" has been developed for the South-East Marine Region (SEMR) already.[52]

Australia was a major original proponent of UNCLOS, and it has enacted legislation to bring its own laws into line with the principles embodied therein. Among the most significant are the Seas and Submerged Lands Act of 1973, the Offshore Constitutional Settlement of 1979, the Coastal Waters (State Powers) Act of 1980, the Continental Shelf (Living Natural Resources) Act of 1968,[53] and the Maritime Legislation Amendment Act of 1994.[54] In addition, although not limited to the marine environment, there is the Environment Protection and Biodiversity Conservation Act of 1999 (EPBC Act),[55] which is somewhat similar to a combination of NEPA and

ESA. The EPBC Act implements CITES, WHC, and other international conventions.

The Seas and Submerged Lands Act[56] basically federalized all of the Australian territorial sea except those state waters that existed prior to federation. It covers territorial sea "baselines" (starting points for the measurement of offshore jurisdictional zones such as the territorial sea), the closing of historic bays, and limits on continental-shelf and territorial sea activities. It also gave effect to some international conventions.

The combined impact of the Offshore Constitutional Settlement and the Coastal Waters Act[57] was to allow the states to make laws governing the ocean adjacent to their territory out to three nautical miles, while leaving all else within the jurisdiction of the commonwealth (the federal government). This recognizes a form of federalism, which permits some local variation on the overall federal theme, along the lines of the U.S. CZMA.

With regard to shipping and ocean dumping, Australia has generally acted consistently with its commitment to UNCLOS. Some of the laws on point are the Navigation Act of 1912,[58] the Protection of the Sea (Prevention of Pollution from Ships) Act of 1983,[59] and the Environment Protection (Sea Dumping) Act of 1981.[60] Each of these, and other legislation as well, addresses aspects of shipping with impacts on the marine environment. Through these laws Australia has sought to implement the key international conventions discussed in chapter two of this book. Ocean dumping, overfishing, spills, and other threats to marine biodiversity are addressed in these laws, as well as under some of the other laws mentioned above.

Notably, under the EPBC Act, Australia has established a Biological Diversity Advisory Committee,[61] with representatives from the government, science, conservation, business, rural, and indigenous communities. The committee advises on conservation and the ecologically sustainable use of biodiversity, including marine biodiversity. It functions as a high-level focal point, with the expertise and broad perspective to point the government in the right direction on big-picture issues. The EPBC Act also has a specific set of provisions aimed at the protection and recovery of threatened marine species and their communities (in sections 248–266A and Part 13), plus special sections devoted to whales and other cetaceans (in sections 224–247).

Australia initiated the Ocean Rescue 2000 program[62] in 1991 to assist the states and northern territory in establishing a system of marine protected areas.[63] The concept is along the lines of Global 200 in that these marine protected areas are to include representative samples of all Australian marine ecosystems. Also, in 1992, the Australia New Zealand Environment and Conservation Council (ANZECC)[64] established a National Advisory Committee on Marine Protected Areas to develop a National Representative System of Marine Protected Areas (NRSMPA),[65] which has since become the Task Force on Marine Protected Areas. The task force has developed a strategic plan to establish the NRSMPA.[66]

Australian MPAs are managed by state or territory authorities in coastal waters, and by commonwealth authorities in areas beyond the coastal waters. All states in the South-East Marine Region have protected areas laws, although only New South Wales and South Australia have specific "marine park" laws.[67]

Of course, one of the crown jewels in Australia's marine environment (or any other) is the Great Barrier Reef, and Australia has taken its responsibilities very seriously in this regard. It has designated the Great Barrier Reef Marine Park (a form of MPA).[68] In addition to the founding statute, there are detailed regulations governing various aspects of aquaculture[69] and prohibitions on mining.[70] The Great Barrier Reef Marine Park Authority (GBRMPA)[71] has oversight responsibilities for all facets of this splendid MPA, and acts as lead agency for WHC issues regarding the reef.[72]

I should note that New Zealand, which is also home to some of the world's most outstanding centers of marine biodiversity, has been considerably less active than Australia in establishing and maintaining its own system of MPAs. The New Zealand Biodiversity Strategy[73] has supported the Department of Conservation in its efforts to add more MPAs, and there is a goal to create several more reserves within the next few years.[74] This initiative was launched in recognition of the fact that New Zealand has lagged behind, with marine reserves protecting only about 0.1 percent of the coastal sea surrounding the North and South Islands.[75]

New Zealand's history is, unfortunately, much more the rule than the exception, and New Zealand is far from the worst offender in terms of laxity in defending marine biodiversity hotspots. Many nations with superlative examples of living marine treasures have done very little to enact and enforce meaningful legislation to regulate overfishing, mining, land-based marine impacts, shipping, ocean dumping, and coastal-zone development.

SOME OTHER NATIONS

Even nations with officially designated MPAs have often been only partially successful in actually protecting them. Inadequate commitment of resources has led to lax or almost nonexistent enforcement of restrictions on harmful activities in and near MPAs. This has resulted in the marine replication of the "paper park" phenomenon that is so familiar from the devastation in terrestrial parks, reserves, preserves, and refuges. One example has been Komodo National Park in Indonesia, where the use of dynamite and cyanide as fishing aids has caused, not surprisingly, tremendous damage to the "protected" coral reefs.[76] Who would have thought? There are ongoing efforts to stem these destructive practices, but it is a daunting challenge given the financial incentives involved and the difficulty of policing large expanses of marine territory. Inasmuch as Indonesia's coral reefs are the most extensive and among the most threatened in Southeast Asia, it is a matter of great concern that they are severely threatened by overfishing, highly destructive fishing (as in the use of dynamite and cyanide), sedimentation, and pollution.[77]

According to a fairly recent World Resources Institute study, out of 646 MPAs in Southeast Asia, the management status could only be determined for 332 of them (not a good sign), and of these 332, only 14 percent were rated as effectively managed.[78] This is highly significant, given that Southeast Asia is considered the global epicenter of marine biodiversity.[79] Its nearly 39,000 square miles of coral reefs, or 34 percent of the world's total, are home to over 600 of the 800 reef-building coral species in the world, about 480 of which can be found in Indonesia.[80] Human activities now threaten approximately 88 percent of the coral reefs throughout Southeast Asia, with the risk to

50 percent of these reefs rated as "high" or "very high."[81] When MPAs, of all places, are besieged by overzealous commercial fishing operations in which explosives and poisons are accepted techniques, the situation is anything but promising.[82]

This sad story is repeated throughout Southeast Asia in country after country, despite extensive national legislation touching on marine biodiversity in most cases.[83] For example, in Cambodia, blast fishing, cyanide fishing, coral collection, trawling, overfishing, and sewage runoff have inflicted much damage to MPAs as well as to less officially protected marine areas.[84] This mirrors the unsatisfactory situation previously alluded to in Indonesia, wherein destructive fishing (explosives, cyanide, and bottom trawling) joins with large-scale land-based pollution, direct mining of coral reefs, and other threats to bring many of Indonesian coral reefs to a "poor" condition, notwithstanding MPA status.[85] Of Indonesia's six Marine National Parks, only three had management plans being implemented as of 2000.[86]

Malaysia has a relatively well-developed system of MPAs, with 64 percent of its coral reefs in "fair" condition, but fisheries remain a threat to East Malaysia while sedimentation jeopardizes West Peninsular Malaysia.[87] Both areas are threatened by dredging, domestic and agricultural pollution, and coastal development.[88] And in Myanmar, the situation appears to be worst of all, with the government actively encouraging and subsidizing the rapid exploitation of natural resources. Dynamite fishing, as well as overfishing in general, and harvesting of live coral and coral skeletons, have caused extensive harm, with no meaningful opposition from the government.[89]

It can be difficult to believe that broad-spectrum poison (usually sodium cyanide) and powerful explosives are actually used in the twenty-first century as accepted fishing practices. But this is the reality, despite national laws outlawing both methods in most if not all of the nations of Southeast Asia.[90] Needless to say, poison and explosives are both extremely crude and indiscriminate methods, killing or injuring many fish and other marine life, including coral reefs, apart from the targeted species.[91] The difficulty of enforcing the existing laws against both poison fishing and blast fishing is exacerbated by the fact that these methods can be used on a primitive level even by small-scale fishing operations. Plastic squirt bottles are filled with crushed sodium cyanide and applied to reefs by divers, stunning or killing the fish and making capture effortless. Similarly, in addition to dynamite and grenades, fishers fill empty beer or soda bottles with potassium nitrate (an artificial fertilizer) and pebbles, topping them with a commercial fuse or blasting cap. When detonated, these primitive bombs kill or injure most of the nearby fish (not to mention reefs and people), causing many fish to float to the surface, while many others sink irretrievably to the bottom.[92]

For the countless fish and coral reefs destroyed by such poisons and explosives, it is small comfort that these deadly tools are officially illegal under the laws of the countries throughout the region.[93] This situation gives a new layer of meaning to the term "epicenter." The waters of Southeast Asia constitute a premier global epicenter of marine biodiversity, and the wanton use of blindly lethal fishing practices helps to make these waters a veritable "ground zero" of attack; they have become the epicenter for the contemporary marine mass extinction. It is as if we had laid out a giant string of red buoys around these irreplaceable reefs as a floating bulls-eye for our ever-continuing, deadly game of target practice.

FOUR

Choosing to Stop Killing Our Oceans

I have shown in the preceding chapters that there is a profusion of law relevant to marine biodiversity. Global and regional international laws aim at various facets of marine environmental health, some much more directly and explicitly than others. These international agreements touch, directly or indirectly, on such topics as ocean dumping, marine protected areas, pollution prevention, preservation of important natural sites, regulation of permissible fishing methods, and restrictions on trade in endangered species. Likewise, individual nations with coastal resources have enacted, one by one, towering piles of statutes governing management of their coastal zones, fisheries, water pollution, ocean dumping, marine protected areas, and protection of endangered marine species.

WHY HAS SO MUCH LAW HAD SO LITTLE EFFECT?

The multitude of laws on many levels is a veritable algal bloom of legislation, a red tide of words. I use these metaphors deliberately, with full awareness of their negative connotations.

As with the terrestrial hotspots, there is a dangerous placebo effect generated by the sheer number and volume of laws that appear applicable. When a layperson, or even a governmental official, sees this many laws all aimed at the same thing, the natural reaction is to presume that all is well. Just look at the names of these legal agreements: "The Convention on Biological Diversity"; "The World Heritage Convention"; "The Law of the Sea Treaty"; "The Marine Protection, Research, and Sanctuaries Act"; "The Marine Mammal Protection Act"; "The Coastal Zone Management Act"; and "The Endangered Species Act." The names sound so promising, so much on point. So many pages of laws, with so many words on each page—they must be effective! All those trees that were felled to make paper to enshrine our cornucopia of legislation could not have died in vain.

The combined placebo effect can anesthetize people, comforting them that the plight of the hotspots has been covered by all these laws. Karl Marx famously opined that religion is the opiate of the masses, but I argue that law has now usurped that dubious honor. Why should people be concerned, much less be galvanized to action, when so many laws from so many sources have already attacked and defeated the threat?

This presumes that people are aware of the current mass extinction crisis and, if aware, that they care. This is probably an erroneous presumption in many cases. I will provide anecdotal evidence in support of this disturbing hypothesis.

I was a speaker at the Annual Conference of the Society of Environmental Journalists (SEJ) in September 2003. The SEJ has a membership of more than one thousand journalists who report on environmental matters, and the conference was well attended. I met with many veteran reporters from the mass news media, including people who work in television, radio, newspapers, and magazines. Repeatedly, the reporters told me the same story: It is very difficult to persuade editors to approve articles and features dealing with our modern mass extinction, because editors tend to believe the extinction crisis is not newsworthy. Senior editors widely consider the mass extinction event to be of little or no interest to their readers, viewers, and listeners.

When I expressed amazement, the reporters stated that editors consider a global extinction spasm to be beyond the radar of their target consumers. Specifically, unless a story has a strong local hook, such as a major employer threatened with bankruptcy or a local health impact, a worldwide mass extinction will seem inconsequential to people interested primarily in matters that affect them immediately and directly.

A similar fate befell a television special dealing with our modern mass extinction. I was one of a handful of people interviewed during the program, the others being uniformly far more famous and far more important than me (including the renowned Edward O. Wilson of Harvard, Stuart Pimm of Duke, and Russell Mittermeier, head of Conservation International). This important show was broadcast all over the world on CNN International. But in the United States it was considered of insufficient interest to viewers, so it was not shown. Why should Americans be uniquely unconcerned about a mass extinction happening right now? Do we, the people of the United States, lead the whole world in our myopic self-absorption and obsession with only those things that directly affect us?

Of course, I have argued in my book *Ark of the Broken Covenant: Protecting the World's Biodiversity Hotspots*,[1] and elsewhere that a mass extinction does indeed affect people in profoundly significant ways, but this message has not gotten through to very many individuals. This apathy may be due in part to the invisibility of many extinctions, extinguishing small, unglamorous, and even unnamed species (often enigmatic microfauna rather than charismatic megafauna) and taking place in remote, inaccessible rain forests and ocean depths. Also, because many extinctions require decades, centuries, or even millennia to become complete, there is no single dramatic short-term headline-grabbing catastrophe to rivet the public's attention at any given point in time. Our current mass extinction does not feature a colossal, big-impact, hit-the-dirt villain such as a speeding mountain-size asteroid. Our villains are numerous, insidious, and widely dispersed, and we tend to focus our short attention spans on more jaw-droppingly immediate disasters.

Public ignorance and apathy concerning our ongoing mass extinction, combined with the placebo lullaby softly sung by choirs of good-sounding laws, has made the plight of Earth's biodiversity a non-issue. The nameless, numberless hosts of species disappearing under our noses and under our waves are as unnoticed or quickly forgotten as information conveyed to the forgetful fish Dory in the film *Finding Nemo.*[2] Such blissful ignorance may be pleasant and comfortable in the short term, but the broader consequences for life on Earth could be devastating. That is why I wrote this book—to provide an antidote to the syndrome of law as the new opiate of the masses.

But if the current aggregation of international and domestic laws has failed to prevent or halt the new mass extinction, both on land and in the oceans, what can be done? Is there an alternative to the trite but untrue formula of more of the same? What better solution is available to the crisis in marine biodiversity?

The easy, and facile, answer is to continue with the mosaic of international and national laws, but to glue in the pieces that have fallen out and replace the Silly Putty foundation with something more solid. Do the key international conventions lack teeth? Then supply them with dentures! Are these treaties ambiguous? Then clarify them! Are they riddled with loopholes? Then close the loops! Are important nations sitting on the sidelines as nonsignatories? Then persuade them to sign! Do the individual nations inadequately safeguard their vital marine resources? Then they should amend their laws and focus on what should be the focal point!

That would be the standard, academic, law review article approach to the spectacular failure of all that law to save marine biodiversity. If so many international and domestic laws have proved utterly inadequate to the task of preventing or halting a mass extinction in Earth's oceans, the default option for a fill-in-the-blanks law review article is to propose amendments to the existing legal framework. Fine-tuning around the edges, a little tinkering with the details here and there, is the paradigm we have come to expect. It is almost as automatic as a colon buried somewhere in the title of a law review article.

I should briefly mention that the conventional analysis would certainly point to some much-needed amendments in the existing international law regime. For example, it would be helpful to amend UNCLOS so that it expressly provided for designation of MPAs throughout the oceans, including the high seas, with a detailed method of selecting sites for protection, explicit standards for the types of activities to be allowed and proscribed within MPAs, and meaningful allocation of resources for policing and enforcement. UNCLOS should also be revised to incorporate a better priority-ranking, eco-system-based system for safeguarding marine biodiversity. Something along the lines of hotspots analysis, Global 200 eco-regions, or WORLDMAP is desperately needed to give meaning to the vague, generic biodiversity exhortations now embedded within the myriad provisions of UNCLOS on multitudinous aspects of marine law.

Similarly, the WHC should be amended to allow world heritage sites (including world heritage in danger) to be inscribed on its lists despite the fact that the sites are situated in areas beyond the territorial sovereignty of any individual nation. The WHC could be more useful with regard to marine hotspots if pockets of endemism were eligible in the high seas areas and all other portions of the ocean outside the grasp of national jurisdiction. If this were accomplished, and if the WHC were amended to provide true enforcement options with con-dign sanctions for noncompliance, that convention could be a pow-erful tool for marine hotspots preservation. All of the above, of course, would have to be effectuated without causing a wholesale exodus of prior signatories from these agreements, while simultaneously at-tracting recalcitrant nations to sign on in the first place. Lots of luck!

Unfortunately, this type of standard analytical approach to legal commentary does not address the fundamental weaknesses of the entire legal structure in the global marine biodiversity context. When the underlying material is rotten, it does very little good to tighten a few loose screws. It is akin to renowned music producer Quincy Jones' description of his attempts to make the songs from the musical play *The Wiz* sound better for the motion picture version: "It's like polishing shit."[3]

This is not to say that there is no value in the current legal regime. In some regions, an individual nation's laws have helped to slow the destruction of important marine hotspots, such as Australia's Great Barrier Reef. Likewise, on occasion a group of nations has come together to coordinate efforts and effectuate local improvements, as with the Mediterranean Action Plan.[4] Because many of the most notable known marine hotspots consist of coral reefs within the EEZ of sovereign nations, there is the potential for very substantial success if more nations were to follow the lead of Australia and New Zealand in aggressively safeguarding these proximal buried/submerged treasures. The WHC could possibly be of significant value in these near-shore hotspots as well, if only the host nations were inclined to inscribe them.

Along those lines, I have already discussed the prospect that international laws such as the WHC could at least get some of the marine hotspots onto the radar (or sonar) screens of significant numbers of people. They have not yet done so, of course. But many near-shore marine hotspots could be inscribed on the World Heritage lists—they definitely satisfy the threshold criteria. And even absent meaningful enforcement provisions, this would awaken some people to the dangers besieging our planet's marine wonders. Awareness of the problem would be a first step toward galvanizing substantive action, a welcome change from the placebo-induced complacency that now cossets us.

There are, however, powerful reasons why international laws have not averted or halted the current mass extinction in our oceans. The forces that impede the noble parade of failed efforts, led by CITES, CBD, UNCLOS, and WHC, are as potent and immutable as those that have conspired tragically for nearly a century to block the Chicago Cubs from a World Series championship.[5] I will enumerate the chief factors that stand as formidable barricades to the international law solution.

Collective myopia must be at or near the top of the list. Individual nations, and their leaders and citizens, are usually very near-sighted when it comes to seeing the forest for the trees, or the ocean for the kelp. They do not see the extraordinary importance of remote marine hotspots to the world as a whole, or to themselves. If the aphorism "Out of sight, out of mind" is true, then nothing could be more beyond the consciousness of most people than undiscovered life forms in the ocean's midnight zone. Just as no light ever penetrates the aphotic zone, no information about deep-ocean biodiversity is visible to us, unless we actively look for it.

A close relative of myopia is a narrow sense of self-interest. Any given nation-state will not sign or ratify an international treaty unless it perceives such a step as furthering its own cause. Genuine altruism is rare among nations, and a treaty will not attract signatories unless there is an apparent advantage to be had by joining. When the United States does not ratify UNCLOS, the CBD, or the Kyoto Protocol on global warming, it refuses to do so because it sees an intolerable disincentive to sign on. A nation that views an international convention as a threat to its economy, whether by mandating the sharing of lucrative information and profits or meaningful reductions in carbon dioxide emissions, will not be a party to such things. As a nonparty, that nation will not be bound, unless the convention is merely codifying what already is in effect as customary international law, and no one can compel a nation to become a party.

Because nations view their self-interest through the lens of their own myopic vision, they usually do not see a particularly robust impetus to sign onto a strong, enforceable treaty that focuses on long-term benefits and geographically remote resources. If the WHC had sharp teeth and powerful substantive requirements, it would not have lured as many signatories as it now possesses. Conversely, more powerful conventions (e.g., the CBD) cannot reel in the United States and other big fish. If species—perhaps yet-undiscovered species—found only in the international waters of the blue ocean might someday offer great benefits to humankind in the form of medicines, genes, or nutrition, that payoff is too speculative, too far off in the indeterminate future, and too diffusely distributed to overcome resistance to such treaties. Without a much more immediate, visible, predictable advantage that redounds directly to a particular nation, the cost-benefit analysis too often works out against becoming a signatory. Something that may be very much in the interests of the whole world lacks sufficient drawing power to pull in key players like the United States.

Ultimately, the lack of a globally recognized court with jurisdiction over international legal disputes, and with the power to enforce compliance with its judgments, supplies the coup de grace for the international law approach. The troubled histories of the World Court/International Court of Justice,[6] and the International Criminal Court[7] illustrate the vertiginous obstacles in the path of progress. As with individual treaties and conventions, powerful nations (e.g., the United

States, again) refuse to acknowledge or consent to the jurisdiction of such supranational judicial bodies, out of concern that submission to their jurisdiction will erode national sovereignty, place citizens at risk, and jeopardize national interests. The United States and some other wealthy, militarily mighty Western nations fear that these courts would be dominated by third-world nations and/or countries that are hostile to them, and would use their powers for political ends without due regard for the rule of law. The specter of politically driven rulings from "kangaroo courts" (or courts named after any other exotic "foreign" animal) is more than enough to frighten these nations away.

Anyone who has ever been on the receiving end of a speeding ticket or a summons and complaint would yearn for this ability to opt out of a court's jurisdiction simply by refusing to cooperate. That is one of the ways in which an individual citizen differs from a sovereign nation-state. A person is, like it or not, compelled by virtue of citizenry or residency to submit to the jurisdiction of all the courts established by the ruling government, at the city, county, state, and federal levels. This person may be philosophically opposed to such jurisdiction, and may be an outspoken critic of the government, but he or she has no ability to walk away once a court with jurisdiction asserts it. The court's jurisdiction is, at the bottom-line level, buttressed with the threat and, if need be, the actuality of physical force against the unwilling subject. Forcible imprisonment and even armed violence stand in the way of a dissident who attempts to opt out of the court's embrace.

Just as individuals cannot decide for themselves which laws will apply to them once they are within the category of persons the government considers bound by the laws, they are powerless to exempt themselves from judicial enforcement of those laws. That is the price we pay for being citizens or residents of a nation or any subdivision thereof, and it is a price that is nonnegotiable. Government is not eBay. If we do not want to do the government's bidding, the only bidding we can do is to pay the price to get out or stay out. If we do not like the laws of a particular nation, we must vote with our feet and physically remove ourselves from that nation's turf. Such is the power of the sovereign nation over the people within its borders. The laws are automatically and universally in force (literally) and of legal effect, by their own terms and on their own terms, without regard for the individualized consent of the governed.

Is force really at the root of law? This hypothesis can easily be evaluated. I recommend that this foray into empiricism be confined to a "thought experiment," rather than an actual test, for reasons that will soon be apparent.[8]

If someone believes herself to be a party to a social compact, where her relationship with the national government is akin to a voluntary association of friends, she can conduct a simple experiment consisting of the following steps: (1) Select one federal law with which she disagrees, such as the income tax. (2) Notify the Internal Revenue Service in writing that she is opting out of the income tax system. (3) Immediately cancel all income tax withholding. (4) Refuse to file an income tax return from this point onward. (5) When contacted by the I.R.S., refer them to the letter submitted under step 2 above. (6) When audited, or when summoned to the I.R.S. offices, politely decline all such invitations, citing prior commitments and the opt-out letter. (7) Continue to decline any government-issued invitations, irrespective of form (summons, subpoena, indictment, etc.).

The result of this thought experiment will be clear to anyone with any experience living in the world on the waking side of our dreams. Are our income tax payments voluntary donations to a worthy cause, such as the contributions we might make to charities, or a gratuity we leave for a service employee in thanks for a job well done? Is our participation in the tax-collection regime optional? If we choose not to comply with the government's invitations to comply, will we be left alone, secure in this exercise of our individual autonomy and personal freedom? No, no, and no.

Once we move past the early stages of this experiment in freedom, it is only a matter of time before form letters and paper persuasion are replaced with the ultimate expressions of government's negotiation skill. Self-addressed envelopes give way to jail cells with steel bars and locks that work from the outside only. Strongly worded letters shuffle aside for federal agents with loaded revolvers and semiautomatic pistols. The extent of our freedom of choice becomes clear as the illusion of voluntariness is supplanted by the actuality of forcible compulsion. We can freely choose to do as the government tells us, and remain out of prison with our tax money safely in the United States Treasury, or we can choose to become intimately acquainted with our new secure location in federal custody (or with the interior of a coffin).

It may be unpleasant to think about the rule of law in this way. For law professors, who inhabit a comfortable world of theory and abstract principles, reality therapy can be strong, bitter-tasting medicine for an unacknowledged malady. But the truth is that within any sovereign nation, the rule of law is buttressed by the use of force. It may not come to that point very often in civilized society, but that is because it is common knowledge as to the ineluctable outcome when people try to defy the government's legal strictures. Remove that ultimate threat of violence, and compliance with the nation's laws would swiftly evolve into what we see so often on the international level: a massive come-as-you-are party where participation is voluntary and obedience is optional.

This is the core reason why international law has not provided and cannot provide the resolution to the mass extinction crisis in the world's oceans. Despite a shipload of voluminous, nice-sounding international conventions, from UNCLOS to WHC, and from CITES to CBD, the legal protection is only as good as the determination and capability of individual nations to do good.[9] Inasmuch as the tangible, direct benefits from taking meaningful steps to preserve marine biodiversity are diffuse and often somewhere in the future (while the benefits from exploiting these resources are immediate and substantial), the results are predictable. They are about as good as what we would expect if the federal government converted the income tax into a purely voluntary program. Suffice it to say that there would no longer be snaking lines of anxious citizens at teeming post offices on the eve of April 15 every year.

If there were a world government, with strong powers over all included governments and peoples, the international system would much more closely mimic the national model. A world government would presumably have a judicial system, including a "world court," that would be the world court in actuality rather than only in theory. This court would have true jurisdiction (read "power") to bring entities before it, and to enforce its judgments. And that enforcement would necessarily include its subordinate component—force. But ever since Alexander the Great brutally united the known world at the point of a phalanx of spears,[10] a forcibly created and maintained world government has not been the dream of most enlightened thinkers.

Idealism, as manifested in a noble faith in the willingness of nations to join together for the common good of the entire world, has spawned such enterprises as the League of Nations and the United Nations. These are tentative steps along the path to a world government. But, in contrast to Alexander the Great's model, these initiatives presuppose a voluntary, cooperative laying aside of age-old hatreds and grievances and an altruistic subjugation of narrow self-interest to the greater cause. To put it cynically, they operate under the delusion that Alexander's bloody phalanxes could be effectively replaced by negotiations, conferences, consciousness-raising, and cooperation. But all the good intentions in the world have proved inadequate to the challenge of slicing through the modern-day Gordian knot of nationalism; narrow self-interest; ancient feuds; religious, racial, ethnic, and cultural divisions; envy; greed; distrust; hatred; fear; and political animosity.

A NEW LEGAL SOLUTION TO OUR MASS EXTINCTION CRISIS

If idealistic vision cannot overcome this thicket of all-too-prevalent human conditions and clear the way for a global commitment to save life in the oceans, what can? I suggest that the only plan with any reasonable prospect of prevailing in this flawed world we occupy must recognize the factors that motivate nations, and turn those factors in the right direction. These motivators might seem ignoble, even base, because they include avarice, selfishness, fear, short-term advantage, and envy. But if people, and the nations that are made up of people, are imperfect and are driven largely by baser instincts, it would be naive and unrealistic to refrain from using these tendencies as means to a more positive end. The "four Ps" of legal realism—*power, politics, purse,* and *prejudice*—can be wielded as tools for progress as well as oppression once we acknowledge that self-interest is what moves the world. As generations of commentators have remarked in many different contexts, "If you can't beat them, join them."

If even one wealthy nation were willing to use its influence, including its money, in the service of marine biodiversity, it could harness the powerful engines of greed and self-interest and put them to work productively. That nation could use debt-for-nature swaps, cash transfers, technology/information sharing, and a variety of diplomatic inducements to encourage other nations to take specific steps to protect the marine hotspots. Optimal selection and preservation of a system of marine protected areas would be at or near the top of the list of desired outcomes.[11] But, although a scientifically sound network of MPAs would be a major and necessary component of the program, it would not be sufficient in and of itself to halt the mass extinction in our oceans.[12] Additionally, nations should be incentivized, even beyond the limits of any MPAs, to eliminate the most destructive commercial fishing practices and the worst methods of offshore exploration, drilling, and mining.[13] Deliberate ocean dumping and proactive measures to reduce the probability and severity of accidental spills would also be targeted. Effective management of coastal zones and land-based activities that affect near-shore habitats would be another focal point.

It is crucial, albeit perhaps counterintuitive, that we pay close attention to land-based activities even as we focus on marine hotspots. There are enormous threats to marine biodiversity that originate, not in the oceans, but on dry land in the coastal zones of the world. Part of the reason these threats are prevalent is that an estimated 67 percent of the entire global human population lives either on the coast or within 37 miles of the coast, and that percentage is increasing.[14] These huge and growing populations often cause overutilization of fishing and other resources in coastal areas, habitat destruction and degradation, pollution (both organic and inorganic), eutrophication and related issues such as pathogenic bacteria and algal toxins, introduction of invasive species, watershed alteration, marine littering, and other harms to the nearby marine regions.[15] Given that so many key marine centers of biodiversity reside in the near-coast coral reefs and continental shelf areas, it is of tremendous importance that our legal approach embrace appropriate controls over these land-based threats.[16] Any plan that shortsightedly and narrowly focuses too much on ocean-based activities will, paradoxically, miss the boat.

Even with regard to land-based pollution, it is a mistake to aim only at direct sources of water pollution, whether from point sources or nonpoint sources. Air pollution can and does often contribute to marine pollution as contaminants eventually settle in the water some distance from shore.[17] Additionally, some persistent organic pollutants (POPs) such as some pesticides and industrial chemicals, radionuclides, trace metals, and persistent toxic substances (PTSs) can be among the most serious chemical threats to the near-shore marine environment, whether they originate in the form of air pollution or otherwise.[18] In conjunction with other land-based pollutants such as hydrocarbon compounds, polycyclic aromatic hydrocarbons (PAHs), sewage, nutrients, sediment mobilization, and litter, these substances can synergistically combine forces to inflict great harm on marine ecosystems.[19] An overarching preservation plan for marine hotspots should therefore take full advantage of the long tentacles of the law to reach land-based activities that significantly impinge on marine biodiversity.

The very serious threat posed by introduction of exotic/invasive species is one that crosses the land/water boundary, both literally and figuratively. Whenever nonindigenous species are artificially introduced into a new habitat, there is grave potential for disruption to the ecosystem. Within the marine environment, invasive species are often brought into new regions inadvertently and unknowingly, as hitchhikers in ships' ballast water.[20] Such species can find highly favorable conditions out of their usual habitat once away from their natural predators, and may out-compete the prior residents, with disastrous results. Any overarching legal plan for the world's oceans must include effective mechanisms to prevent further introductions of exotics, especially in key marine hotspots.

All of the above issues notwithstanding, a large network of well-chosen and zealously guarded marine protected areas is perhaps the most indispensable ingredient in any effective legal response to the threats to life in our oceans. Whether near a coastal zone or not, MPAs must be protected (in more than name only) under this plan. Paper parks, whether on land or in the oceans, are worse than useless because they can deceive us into believing that the problem has been solved. But intelligently chosen, appropriately sized, and vigorously regulated MPAs have been proven to be effective in preserving and replenishing populations of threatened marine species.[21] There is an entire field of study devoted to effective selection, design, and management of MPAs, and it is beyond the scope of this book for me to attempt to go into detail on its tenets here.[22] There is some controversy regarding the optimal location choice for MPAs, for example, both on political and scientific grounds, but there is much more consensus on the scientific criteria than on the economic, social, and political practicalities that often collide with scientific factors.[23] Where there is sufficient information to determine the location of a marine hotspot, for example, there is little doubt that most scientists would concur on the advisability of creating a refuge in that place; but that may not be feasible given all of the extra-scientific realities that must also be dealt with.[24]

There are legitimate technical/scientific issues regarding the best design of MPAs as well, in terms of such parameters as size and number,[25] connectedness to other reserves or to ecologically important ocean currents, and the extent to which MPAs should be open to "multiple uses" aside from strict and exclusive conservation.[26] At present, there are many different types of MPAs, with a wide spectrum of activities permitted and degree of protection afforded.[27] Just as there are many types of terrestrial protected areas, including wilderness areas, wildlife refuges, national parks, and national forests, there is a menu of options available under the rubric of MPAs. The option that would probably be most appropriate for marine hotspots, because it is most exclusive of nonpreservationist uses, is the marine reserve, wherein no extractive use of any resource (living, fossil, or mineral), nor any habitat destruction is allowed.[28] Additionally, many MPAs are designed with a type of zoning, which permits different practices according to the part of the MPA that is in question.[29] But there are difficult lines to draw in the sand, as external pressures argue for less-strict forms of MPAs, such as seasonal closures, bans on taking only reproductive individual specimens, moderate catch limits, restrictions but not prohibitions on mineral extraction, limitations but not bans on certain types of fishing methods (such as trawling), and regulation of waste disposal.[30] Every time a compromise is forged on such issues, a crack opens in the shield around marine hotspots, through which the multitudinous seahorsemen of the apocalypse can enter and pillage these fragile, vital centers of oceanic endemism.[31]

Moreover, the job is far from over once MPAs are selected and delimited; there remain serious and long-term choices to be made with respect to monitoring, policing, further research, proper management of areas near or adjacent to MPAs, and other concerns.[32] It is vital, for example, to ensure that MPAs are not harmed by pollution, overfishing,[33] runoff, and other activities that take place beyond the MPAs themselves.[34] Thorough planning and continuous, flexible, interdisciplinary management are essential to a successful MPA.[35]

Some recent important studies have listed ten major criteria that should be considered with regard to management choices for any MPA or marine reserve.[36] The specific needs of each marine area are different, reflecting the varying degree and types of threats, multifarious physical and biological features, and other variables, and thus there can be no single "correct answer" to the question of when and how to implement marine protected areas.[37] These criteria are not necessarily to be weighted equally, but each is significant to some extent:

1. Biogeographic representation. It is desirable to include within the MPA or network of MPAs representatives of as many different biogeographic zones as possible.
2. Habitat representation and heterogeneity. MPAs should be chosen so as to include examples of all different marine habitat types. This is consistent with the Global 200 Ecoregions approach to setting conservation priorities.
3. Human threats. There is a need to protect reserves from indirect or nonextractive human impacts, such as pollution, runoff, and habitat alteration.
4. Natural catastrophes. Whether a particular reserve area is subject to severe natural catastrophes should be taken into account.
5. Size. An MPA must be of sufficient size to meet its goals, capacious enough to supply adequate territory to all the species it is intended to safeguard.

6. Connectivity. A reserve's connection by dispersal to other reserves or to the rest of the ecosystem is an important factor in determining its overall efficacy. Isolated reserves are generally not as effective as those that are connected.

7. Vulnerable habitats, life stages, or populations. These at-risk entities are in particular need of MPA safeguards.

8. Species of particular concern. Endangered and threatened species, such as those specified under the CITES paradigm, are likewise appropriate beneficiaries of reserve protections.

9. Exploitable species. When a reserve is home to commercially valuable species that are exploited outside the reserve, this should be factored into the determination of the size and allowable activities of a reserve.

10. Ecological services for humans. If the MPA provides substantial ecosystem services of benefit to people, that is an additional reason to safeguard it.[38]

These are all precisely the types of issues that should be intelligently addressed, nation by nation, in a cooperative fashion under the auspices of a single nation's statute that provides both practical bottom-line impetus for action and the scientific and technological resources to make effective conservation attainable.[39] Such cooperation currently is quite rare, about as common as the leaders of all the world's nations strolling into a flower-filled meadow while holding hands and reciting in unison the recovering-predator sharks' slogan, "Fish are friends, not food."[40] But it need not be a fairy tale if we use the legal tools that recognize the reality of human motivations rather than those that pretend to ignore them. Multinational synergy could be taking place right now to a far greater extent than it is, if only the disincentives were supplanted by positive inducements.

Why should this daunting challenge be appropriate for one nation, acting alone, to take on with its own federal statute? If, as I have shown, the conventional international law approach has not worked and truly cannot work in the real world, is there any reason to think that one nation could step into the void and supply the legal impetus for a global sea change in how the world's nations treat the living things in the oceans? Surprisingly, the answer is yes. Just as surprisingly, given our record as a nonsignatory to key international environmental agreements, the United States has led the way.

Consider a federal statute that identifies key regions within other nations (or in international waters) that are very important, but that are being destroyed under the status quo. The statute would set forth a mechanism for intelligently selecting appropriate marine areas for a variety of MPA categories, with substantial infusion of international and local technical and scientific expertise as to the location, size, and permissible activities in and near each specific MPA. The statute would fill a great need worldwide by offering a rigorous information production and dissemination framework for targeting the marine eco-regions most in need of extraordinary safeguards, whether by virtue of a high endemism rate, elevated risk of serious damage from human activities, or unusual/unique habitat features. Once identified, these sites would then be evaluated individually by an international/local panel of experts to determine the optimal menu of protection options needed to sustain the site's biodiversity in the long term, and the range of human actions that should be allowed or curtailed in support of that goal.

As I have discussed, MPAs are not a once-size-fits-all phenomenon, and the proposed statute should be aimed at developing a site-specific set of recommendations based on all relevant factors. Some marine regions would require more wide-ranging and stringent protections than others, based on degree of threat, size of the key area, and level of importance of the biodiversity therein, so the law should allow for a multitiered array of MPA options, along the lines already in place under various legal regimes. It would also be crucial for the statute to provide a holistic approach to these MPAs so that the sites are not picked in isolation, but rather with an eye toward establishing and maintaining a reasonably comprehensive global network of marine protected areas, representative of all key marine habitats and ecosystems, with enclaves for all known and probable centers of marine endemism. As with terrestrial reserves, it may be important to choose protected areas in multiple locations, with natural avenues of connectivity, such as through major ocean currents. My proposed statute can supply this type of big-picture perspective because its charter will be crafted with an overarching objective in mind, rather than a piecemeal, nation-specific focus.

Such a statute could be the vehicle that disseminates information about these vital areas to all other nations. Even if that were its only

contribution, it would still be worthwhile, because at present there is a
dearth of reliable information on marine biodiversity and important
regions for conservation priorities on a government-to-government
level. This type of information-sharing function is one of the best
features of the WHC, but, unlike the WHC, my proposed statute
would not be artificially constrained against focusing attention on
areas not within the territorial sovereignty of any nation. The statute
would be able to identify for heightened conservation efforts any
scientifically worthy site anywhere in the Earth's oceans, whether on
the high seas or within the territorial waters of a particular nation.
It would be as flexible and versatile as called for by the evolving state
of the scientific information available to support a site for special
protection.

The statute would also, more generally, establish scientifically
supportable limitations or bans on various forms of trawling, dredging, use of drift nets, ocean dumping, marine mining and exploration,
and coastal zone activities on a situation-specific basis. This portion of
the act would probably focus primarily on these activities in relatively
close proximity to the marine hotspots and the MPAs that would be
established to encompass them, although it would also be useful to
discourage such harmful enterprises in areas more geographically
removed from the hotspots, because these stressors tend to imperil
marine biodiversity wherever it is situated. Left unchecked, such
practices could create more endangered ecosystems and more imperiled species, so it would be proactive and prudent to address them
before they cause further crises.

In this regard, the proposed statute would be capable of directing
attention to all of the destructive fishing practices described in this
book, with creation of appropriate guidelines for the elimination or
regulation of each of them, either on a global level or only within
certain vulnerable marine regions; the same is true of marine exploration and extraction of oil, gas, and other valuable commodities.
Indeed, for any significant threat to marine biodiversity, the statute
would provide the legal framework for the formulation of reasonable
standards that would ameliorate the harm being inflicted. Again,
under the auspices of the enacting nation, these decisions could and
should be informed by regular, systematic, and significant levels of

information sharing and debate with other nations, especially those most affected by any given rule.

On a practical level, the statute would work by providing tangible inducements for other nations to take appropriate steps, via debt restructuring/forgiveness, outright cash grants, and other forms of financial aid. These incentives would be available to those nations that qualify for them by virtue of verifiable actions taken by them to preserve the MPAs and comply with guidelines established under the statute governing other activities affecting centers of marine biodiversity (destructive fishing practices, marine pollution, harmful mining activities, excessive coastal runoff, etc.). I have proposed this type of unilateral, inducements-oriented statute with regard to the terrestrial hotspots,[41] and the concept is perhaps even more appropriate for the largely international realm of the marine hotspots, where many sites are beyond the territorial sovereignty of any single nation. Thus, most of the activities the statute would seek to influence take place in international waters, diminishing the extent to which people might object to the proposal as an intrusion upon the private domestic/internal affairs of other countries.

An incentives-based statutory approach would be deferential toward national sovereignty while still offering a utilitarian, tangible, immediate motivator for each nation to decide to take appropriate actions. It would not purport to mandate or force any nation to take any particular actions or to refrain from any specific activities, only to give nations some attractive incentives to follow proposed guidelines voluntarily. Instead of coercion, it would offer expert information and an array of attractive reasons why nations should opt to cooperate with a biodiversity-friendly course of action. Rather than arrogating to, for example, the United States the role of world eco-cop, it would establish this nation as the leader and primary financial backer of a movement to steer the world in a more responsible direction regarding the most global of global resources.

Among the important advantages of this approach is the issue of practicality (i.e., the degree of difficulty associated with enactment in the first place). On a threshold level, the proposed statute would only need to attract the support of a simple majority of both houses of Congress and the president. That may not sound like an easy task, and it often is not, but it is considerably more feasible than garnering the support of the United Nations and navigating an international treaty through the straits erected by nations that are hard-wired to oppose anything the United States supports. The divide between Republicans and Democrats in Congress is dwarfed by the chasm between this nation and many others, including erstwhile "allies" such as France and Germany as well as a host of more overtly hostile countries. And, if the international convention option is employed, there will always be the problem of nonapplicability to nonsignatories. The job will never really be done until all nations are on board, whereas under my proposal the statute and its benefits will be instantly available to every country upon enactment.

Plus, as I mentioned, there is in fact recent evidence that the United States recognizes both the need for and the possibility of this country taking the lead internationally with this type of action-spurring, inducement-based legislation. In enacting the Tropical Forest Conservation Act of 1998,[42] Congress determined that the United States should protect tropical forests because they benefit humankind through biodiversity, agricultural resources, balancing global climate, and regulating hydroelectric cycles; Congress recognized that one of the causes of rampant deforestation is the enormous debt load some poorer countries carry, which impels them to exploit their tropical forest resources.[43] I will discuss this in some detail to clarify how the concept could be transplanted into a similar statute aimed at safeguarding the world's marine hotspots.

The Tropical Forest Conservation Act is intended to protect tropical forests by alleviating debt in qualifying countries, and to target money for the protection of tropical forests using "debt for nature swaps." Although hampered by numerous qualifications unrelated to biodiversity issues,[44] dependent on continuing appropriations of necessary and meaningful amounts of debt-forgiveness funds by Congress, and largely left to the discretion of the president of the United States, this statute is at least a step in the right direction. It stands as proof that the United States is aware of both the value of global biodiversity and the power of economic incentives to drive appropriate remedial measures in other sovereign nations, just as economic conditions have been a powerful force driving poorer nations to overexploit their natural resources.[45] It is encouraging to note that the 1998 act was overwhelmingly approved by the House and passed the Senate under unanimous consent.

The Tropical Forest Conservation Act authorizes the president to allow eligible countries to use debt swaps, buy-backs,[46] or debt reduction/restructuring[47] in exchange for protecting specified threatened tropical forests on a sustained basis. The president can use the act to reduce some bilateral government-to-government debt owed to the United States under the Foreign Assistance Act of 1981 or Title I of the Agricultural Trade Development and Assistance Act of 1954, or to restructure debt to an amount equal to or lower than its asset value. The secretary of state is empowered to negotiate these bilateral agreements. In return, each of the recipient nations is to put its own money (in local currency, as opposed to the usually required hard currency) into a tropical forest fund to pay for preservation, restoration, and maintenance of its forests. The act allows private organizations and NGOs to contribute their funds as well, in what are called "three-party swaps."[48]

The Tropical Forest Conservation Act attempts to ensure accountability through establishment of an administrative body within each beneficiary country. This group is to consist of one or more U.S. government officials, one or more persons appointed by the recipient country's government, and representatives of environmental, community development, scientific, academic, and forestry organizations of the beneficiary country. These groups are all overseen by the pre-existing Enterprise for Americas Initiative Board, which was expanded by four new members under the act.

The act was reauthorized in 2001, through Fiscal Year 2004, and again in September 2004 for an additional three years—an indication of some initial successes, the continuing support of Congress, and the endorsement of President George W. Bush.[49] The first actual debt-for-nature agreement under the act was concluded in 2000 with Bangladesh; Belize, Thailand, and El Salvador followed close behind in 2001. Several other nations have since followed suit with their own agreements. Between the time of the act's enactment and February 2002, $24.8 million had been used to restructure loan agreements in four countries.[50] The reauthorized act had appropriations of $50 million, $75 million, and $100 million for Fiscal Years 2002, 2003, and 2004 respectively—a sizable increase from the $13 million appropriated for both Fiscal Years 2000 and 2001.[51] Unfortunately, funding levels have now receded a bit under the latest reauthorization, with $20 million, $25 million, and $30 million appropriated for debt reduction in Fiscal Years 2005, 2006, and 2007 in turn.

Of course, this single example does not prove that the United States, or any other prosperous nation, would be willing to establish a similar act to save marine biodiversity. Because numerous key areas are in international waters, any effort to incentivize nations to protect them could not be aimed only at one nation, but rather at *all* nations with a significant history of causing problems or with the ability to begin doing so. If we were, in effect, to pay nations to respect a system of MPAs and not to overfish, employ drift nets, use trawls or dredges, or dump pollutants in or near these waters, we could be inadvertently establishing a perverse incentive for more nations to begin harming the hotspots, if only to qualify for "protection money" later on. And because such vast areas would be covered by marine protected areas and other restrictions, there would be very formidable challenges regarding monitoring and enforcement. The benefactor nation (read the United States) would need to have a basis for assessing whether the recipient nations are really complying with their part of the bargain. This would be no easy feat given the vastness and depth of the target oceans.

There are sizable costs associated with any comprehensive initiative to establish and maintain a scientifically sound system of marine protected areas coincident with the marine hotspots. The nations that forego fishing/trawling, mining, dumping, coastal zone pollution, and other harmful activities in and near these MPAs would sustain significant lost opportunity costs. If a benefactor nation seeks to replace these losses as part of a plan to incentivize other nations to behave responsibly, this would add a similarly substantial amount to the benefactor's tax burden.[52] However, there are also offsetting values gained from preserving MPAs; these range from biocentric intangible values derived from doing the ethically right thing[53] to more "cash on the barrelhead" economic values that accrue from new discoveries, larger available populations of commercially important species (including fish) beyond the MPA, and enhanced ecosystem services.[54] These gains can both assist in incentivizing nations not to harm the MPAs and in compensating the benefactor nation for some of the costs it incurs in implementing its legislation.

This hotspots preservation program need not be entirely the province of government. There can and should be a major role under the statute for various NGOs with regard to supplemental funding, educational and public awareness initiatives, policy and technical guidance, and liaison with the affected communities. NGOs can be especially effective in spearheading the community-based aspects of MPA selection, establishment, and management, because they have the flexibility and local expertise that is often lacking within governmental bureaucracies. By contributing specialized knowledge and experience, NGOs could be instrumental in assisting with the optimal siting decisions as well as in determining the mix of activities to permit and disallow in and near the individual MPAs in each case.

With the right set of incentives in place, NGOs and governments can cooperate with the regulated people and work toward a natural resource partnership.[55] This is vital, because without local buy-in from affected communities (fishing industry professionals, indigenous hunters/fishers, coastal zone farmers, cruise ship personnel, shipping industry people, etc.) any new effort to protect hotspots through MPAs will soon become an old-fashioned, top-down, trickle-down scheme, as familiar as it is ineffective.

As with the local administrative bodies established under the Tropical Forest Conservation Act, I envision the marine hotspots statute providing for a U.S. expert-level group presiding over and coordinating the efforts of organizations within each nation significantly affected by the act. There should be representatives of the U.S. government, the local nation's government, prominent interested NGOs, and specialized subject-matter experts (marine biologists, fisheries specialists, ecologists, etc.) in each organization set up under the statute. In the host nation, opportunities for public notice and comment regarding important decisions would add to the credibility of each local board's decisions, and community outreach would be instrumental in educating and persuading people "on the ground" and in the water, nation by nation, who would need to live with the regulations established.

These local administrative boards would serve as mechanisms to ensure that competent scientific and technical opinion receives a fair hearing when decisions are made about how to protect marine hotspots. We can rest assured that political and economic concerns will be given full voice, as they always are in legal matters (and perhaps ineluctably in all human enterprises, however noble and lofty the aims). By mandating that certain membership slots on the boards be filled by people of appropriate expert qualifications, and by expressly allowing the boards to consider input from concerned citizens and outside organizations, the proposed statute can supply the opportunity for regular, routinized, and continual injections of sound science into the debate. Politics and purse may still predominate, human nature being what it is, but they can be informed by solid facts and scientific principles. And that may be the most we can ask of any process run by human beings.

The boards, with their embedded scientific and technological expertise, would be more than a bureaucratic, policy-formulating construct. Under the statute, they would also be empowered to devote resources and attention, on a priority basis, to sectors of the marine environment deserving further study and research. As discussed in previous chapters, we have much to learn about phenomena such as seamounts, hydrothermal vents, deep-sea benthic ecosystems, and the largely uncharted diversity of life among demersal species. Outstanding known examples of such natural treasures would certainly be eligible for the boards to select for inclusion within new or existing MPAs, but prudently targeted research efforts can be expected to yield a trove of new discoveries and new information as well. Working in cooperation with NGOs, universities, and scientific organizations, the boards would serve to expand the horizons of human knowledge. By thus examining the oceans on a systematic, conservation-oriented level, we can begin to clear up some of the numberless mysteries that have for so long obscured so much of this planet. Some of this new knowledge will ultimately lead to noteworthy advancements in both applied and general science, in fields such as medicine, agriculture, biotechnology, genetics, evolutionary biology, ecology, and chemistry.

One factor in my proposal's favor is that there are comparatively few nations with significant commercial fishing or shipping presences in the open ocean, or even the potential to acquire them, which would narrow the list of nations with whom to negotiate over MPA safe-guards, regulations on destructive fishing/mining, and ocean dumping. And for the nations with major coastal zone/continental shelf resources, the geographic areas in question are of more manageable size, and much closer to shore, rendering monitoring and enforcement more feasible. Some of these nations are already behaving responsibly; right now they have very little, if any, self-interested incentive to be good stewards of the environment, so presumably they would not begin doing otherwise in hopes of extracting some payoffs.

It is important to note that an incentives-based galvanizing statute would not require any forays into such controversial and legally dubious notions as the expansion of "universal jurisdiction" in order to be effective. I am certainly not advocating that the United States join the ranks of nations that are now vigorously asserting universal jurisdiction over foreign dictators, war criminals, and other disfavored people from other lands in a misguided attempt to bring them to justice in their own courts.[56] In my view, this extension of the venerable concept of universal jurisdiction is a blatant and dangerous power grab, without sound basis in international law. It is an effort to arrogate to an individual nation the power to coercively apply its legal system to anyone it targets, irrespective of a person's citizenship, and to subject individuals to the nation's version of justice within its own judicial system. It is an acutely political notion, driven by political passions and prejudices, and facilitated by the elastic concepts of war crimes and crimes against humanity. In contrast, my proposal is consistent with established international law, and entails no assertion of civil or criminal jurisdiction over citizens of other nations, nor any other form of force or coercion.

It is significant that the proposed statute should not run afoul of countervailing legal regimes such as the General Agreement to Tariffs and Trade/World Trade Organization (GATT/WTO) either.[57] The difficult issue of environmentally motivated trade sanctions and the extent to which they can be held violative of international free-trade principles should not apply to the situation in which one nation offers a benefit to other nations in exchange for adjustments in behavior. In contrast to punitive economic measures, these positive incentives would not penalize a nation via discriminatory restraints on free trade. Although financial advantages would certainly flow from debt restructuring or forgiveness or cash transfers, these "carrots" should not be elided with "sticks" such as revocation of most-favored-nation status, imposition of punitive tariffs, or the like.

Moreover, the very existence of, for example, a U.S. statute along the lines of the Tropical Forest Conservation Act aimed at identifying and protecting marine hotspots, should be helpful in dissuading some nations from continuing their destructive practices. If a well-established, developed nation such as Japan or Russia is engaged in negotiations because of irresponsible marine activities, and the prospect is raised that they could be offered financial inducements to stop, they might be moved to implement reforms out of a sense of shame and/or the global equivalent of peer pressure. The statute would shine a bright spotlight on nations that fail to comply with scientifically robust guidelines regarding MPAs and harmful ocean practices, and it may be that the court of public opinion, worldwide, would be as effective as the lure of financial gain in motivating some nations to choose compliance over defiance.

PROOF THAT A NEW LEGAL EFFORT IS WORTH IT

Is my proposal hopelessly utopian, if not contrary to law? Could or would the United States build on its successes with the Tropical Forest Conservation Act, accept this new challenge, and become the global leader in marine biodiversity preservation by enacting and implementing another incentives-based federal statute focused on this goal? Given our fetid record regarding UNCLOS, will we now perform an about-face and take the lead in global marine protection, in a surprise O. Henry ending? Is this even the best way to deal with the situation?

I am acutely aware that this proposal runs counter to some deeply held beliefs in the primacy and efficacy of international law in such global matters. I know that a U.S. statute is anything but the accepted approach to this type of international issue, and that many see the evil of imperialism barely concealed behind the veneer of this method's altruism. Distilled to its core essence, my response to these objections is, to phrase it colloquially: What part of "mass extinction" don't you understand?

If the international law system were an adequate preventative or panacea for the plight of marine biodiversity, why has the first mass extinction since the K-T spasm erupted in the midst of all that legislation?[58] Certainly it cannot be that the members of the United Nations are agnostic or ignorant to the importance of preserving biodiversity and the many synergistic threats to it,[59] or that they lack the resolve to try to do something about it. International conventions from CBD to CITES and from UNCLOS to Bonn stand as evidence to the contrary. Given the amount of attention the international community has devoted to marine biodiversity, it is clear that this community has made its best effort, over a span of decades, to solve the problem through traditional international legal means. If this method could prevent or halt the mass extinction, it would have done so. It has not done so, because it cannot. As the saying goes, "I would if I could but I can't so I won't." The problem with international law as the solution to our mass extinction is not that we need more conferences, more negotiation, more noble resolutions, more fine-tuning of the existing conventions, and more time for the system to work. It goes much deeper than that, to the heart of the whole system.

The flaws in the international law approach, as discussed in this book, are as intractable and interconnected as they are numerous and ubiquitous. When we attempt to rectify any one of them, we ineluctably exacerbate another. It is the legal equivalent of trying to get rid of dirt by sweeping it under a carpet. Once the dirt (i.e., the set of drawbacks inherent in international law) is under the carpet, any effort to make it disappear by stepping on an unsightly bulge only results in that bulge migrating to another section of the rug. It is almost akin to a law of physics: dirt under a carpet can neither be created nor destroyed, only moved. If a treaty cannot garner enough signatories, the treaty is weakened with exceptions and reservations to render it more attractive to reluctant nations. If a treaty lacks enforcement mechanisms and therefore cannot compel its signatories to comply, any attempt to remedy this deficiency will likely cost it a significant number of parties. And so on. Put pressure on one problem area, and that pressure shifts elsewhere, causing a different trouble spot.

The fixation of multitudinous legal commentators on an international law paradigm for all extraterritorial challenges, despite mountains of failed treaties lining the road to the first mass extinction in 65 million years, is understandable. International law is the comfortable, well-accepted, politically correct, standard answer. It requires no unsettling confrontations with the reality of our decades of dismal legal experiments at the expense of our planet's life. Indeed, this penchant for the familiar, mail-it-in approach is reminiscent of the phenomenon Eric Hoffer observed in human nature, wherein so many people prefer a good alibi to genuine achievement.[60] Within the international law regime, the alibis are well known and built into the system. All knowledgeable persons know the alibis are there, and they accept them, because international law is the One True Answer, and the alibis simply come with the territory.

As much as the concept of international law might appeal to us on a philosophical level as the optimal, if not the only, appropriate mode of handling global issues, there is no efficacious means of remedying the practical difficulties built into the international law system. To return to the analogy, once the dirt is swept under the carpet, no amount of dancing around on top of the problem is going to remove it. The only real solution is to lift up the rug and get at the dirt directly, even if we soil our nice, clean hands a little in the process. In other words, break out of the international law paradigm and try a completely new approach to come at the problem from another direction. That is my proposal for an incentives-based U.S. statute aimed at the marine hotspots.

In my book *Ark of the Broken Covenant*, I argue that even a nation motivated mostly (exclusively?) by narrow self-interest and greed should rationally determine that it is wise to commit significant resources to preserving hotspots, if all the relevant factors are weighed logically. I call the analytical framework that yields this result the Hotspots Wager, and the corresponding Decision Matrix illustrates the outcomes from all the various possible combinations of variable values.[61]

The Hotspots Wager is a useful method of rationally assessing the optimal decision nations should make, given the multiple and enormous unknowns inherent in the hotspots concept. There are vast unknown and unknowable gaps in the pertinent facts relevant to the oceanic realm (even more so than in the case of the terrestrial hotspots) that any rational utility maximizer would want to know when determining a course of action.[62] This constitutes a formidable epistemological puzzle: How do we know what we do not know?[63] It is tremendously important that we look at the missing links in our information chain the correct way, lest we make the dreadful mistake of assuming that all those question marks mean we should take no action.

In simplified form, and as applied specifically to marine biodiversity, these great unknowns are three in number:

1. How many species actually exist in the marine hotspots now, including all that are currently unknown to science?
2. What is the true tangible value of these species to humankind, both now and in the future, including benefits people derive from ecosystem services?
3. How great is the actual extinction risk for the species in these marine hotspots, on average?

By comparing the magnitude of the consequences that follow from each possible combination of potential extreme values for each of the three main variables, the Decision Matrix allows us to conceptualize the benefits and risks inherent in any determination to fund or refrain from funding a major program to preserve the marine hotspots. In essence, the Hotspots Wager is a gamble where the stakes are extraordinarily high, and where the decision-makers must find a reasonable way of dealing with at least these three huge unknown factors. The Decision Matrix places the likely outcomes from each combination of extreme values for each variable in juxtaposition with one another to allow us to evaluate whether there is a greater risk from nonaction or a greater reward, and vice versa.

How much money are we talking about wagering here? That is, how much money would it cost to implement a proposal along the lines I am advocating, so that we can understand the amount of dollars hanging in the balance? Certainly there would be economic benefits as well as costs if we were to fund a reasonably adequate global network of marine protected areas, and jobs would be created as well as lost, so such calculations are not a simple matter. However, there have been some credible attempts to arrive at an estimate of the net costs of a representative worldwide system of MPAs. One recent estimate is an annual outlay between $5 billion and $19 billion, although this cost perhaps could be considerably reduced by the elimination of most government subsidies to destructive marine activities that would be unprofitable but for the government handouts.[64] Also, among the considerable rewards would be substantially healthier and more sustainable fisheries, plus more reliable ecosystem services of vast value.[65] For nearby land masses as well, especially islands, there can be a real plus side to the ledger when MPAs are created and effectively protected off their shores.[66]

With billions of dollars at stake, and gigantic consequences possible from certain particularly wise or unwise decisions, how should we choose what to do about hotspots preservation? The following table, the Decision Matrix, is my attempt to simplify the main issues relevant to the question of whether an incentives-based legal solution to the hotspots puzzle should be implemented. The table distills the primary question marks in the hotspots equation into the three unknowns (that may well never become known), as previously mentioned. These three unknowns form the core of most of the objections to my approach outlined above. Critics would argue that these unknowns probably cannot be ascertained, and that in light of so much uncertainty it would be irresponsible and imprudent to risk billions of tax dollars a year on safeguarding hotspots. Are they right? The Decision Matrix can help us decide.

Obviously, the table is intended as a simplification. I recognize that the true situation as to each unknown, if we could somehow determine it, would be some complex and shifting position along a continuum of possibilities. Because we are lumping together *all* of the countless species in the oceans, this subsumes immense variation from species to species on each of the variables. Nevertheless, for purposes of framing the issues, I have boiled down the value of these unknowns to two polar opposites at the extremes of each continuum, either "low" or "high." How's that for oversimplification? I have reduced the most momentous questions in all of biological sciences to the binary values of "zero" or "one." But there is a good reason for this. Anything between these limits would merely be variations on the general theme.

Within a given variable, there can also be complicating factors. For instance, some species are at much higher extinction risk than others within any hotspot; some hotspots as a whole are in greater danger than others, and/or would cost more to preserve; some species have much more current or future practical value than others; and some hotspots contain far greater numbers of species and/or more valuable species than others. Also, much of the practical value of a particular hotspot could theoretically be confined to one species among the hundreds of thousands that reside therein. Such factors as these could and should be used to craft individually tailored regulations and management plans for each hotspot under the legislation I envision, but they need not detract from our use of the Decision Matrix as an illustrative tool to shape our more general decision-making.

One other point deserves explanation. The variable for practical value of all species within hotspots encompasses both identified and unidentified species. It also includes both currently known uses and those that still wait to be discovered or needed. It may be centuries before we learn about certain benefits we could derive some a particular species' genotype or phenotype. Plus, new diseases, new environmental stressors, changed atmospheric conditions, and other unpredictable future events could be many years away at present, but someday they may confront us, and a previously "insignificant" species could suddenly take on great value by offering the solution. I could have designed the Decision Matrix with separate columns for current and future value of species, and/or for known and unknown species, but this would have complicated the table without real gain in utility. The appropriate decisions as to hotspot preservation would not be altered much, if at all, by separating the categories of species value in this manner, so I have placed them in one variable.

The "Results" column represents the principal types of consequences that flow from a decision about whether or not to invest heavily in hotspots preservation, depending upon all possible combinations of the value of the three unknowns. There are eight different ways in which the "high" or "low" value of three unknowns can be combined, and those eight combinations yield some dramatically different results. I have used very abbreviated shorthand labels to describe the various possible results, along the lines that might be used in game theory or in analyzing a game of chance in which wagers are placed. I use the terms "First Order" and "Second Order" to denote respectively, in broad terms, the more significant and less significant variants within a particular category of impact. I might just as well have chosen the plain-English words "big" and "small" but that would not have sounded as impressive or as academically erudite, so I opted for pretentiousness. I am, after all, a law professor.

Enact and Fund Major Hotspots Protection?	True Degree of Extinction Risk in Hotspots	True Number of Unknown Species in Hotspots	True Tangible Value of All Species in Hotspots	Results of Funding Decision
No	Low	Low	Low	Lucky Wager, Money Saved
No	High	Low	Low	Second Order Serious Error
No	Low	High	Low	Lucky Wager, Money Saved
No	High	High	Low	First Order Serious Error
No	Low	Low	High	Lucky Wager, Money Saved
No	High	Low	High	Second Order Grave Error
No	Low	High	High	Lucky Wager, Money Saved
No	High	High	High	First Order Grave Error
Yes	Low	Low	Low	Unused Insurance
Yes	High	Low	Low	Second Order Soft Benefit

Enact and Fund Major Hotspots Protection?	True Degree of Extinction Risk in Hotspots	True Number of Unknown Species in Hotspots	True Tangible Value of All Species in Hotspots	Results of Funding Decision
Yes	Low	High	Low	Unused Insurance
Yes	High	High	Low	First Order Soft Benefit
Yes	Low	Low	High	Unused Insurance
Yes	High	Low	High	Second Order Jackpot
Yes	Low	High	High	Unused Insurance
Yes	High	High	High	First Order Jackpot

Let me explain the bad news outcomes first. A "Serious Error" is a failure to protect the marine hotspots when there is in fact a major extinction risk for the species therein but the tangible value of those species overall is low. This is a *serious* and not inconsequential error because presumably some species will go extinct due to our inaction, and they will have at least intangible value. If there are many unknown species, this value is multiplied greatly, resulting in a "First Order Serious Error," while if the number of unknown species is actually low, we have a low multiplier effect and a "Second Order Serious Error."

Similarly, a "Grave Error" is a failure to protect marine hotspots when there is in fact both a major extinction risk for whatever number of species live therein and a high tangible value for those species. This is a *grave* error because some species will die out that could have provided people or the planet with great benefits, such as cures for disease, valuable genes, ecosystem services, new sources of nutrition, and other benefits.

The accelerating and potentially catastrophic loss of biodiversity is different in kind and not only in degree from all other environmental threats because once a species is committed to extinction the harm is irreversible. That is why I chose the term "Grave Error." Unlike air pollution, water pollution, toxic-waste dumping, or any other form of environmental harm, the destruction of entire categories of life is a wrong of staggering proportions that can never be righted no matter how much money we throw at it and no matter how hard we try. Once the living product of millions of years of refinement is shattered into extinction, no subsequent penalties on those who caused it, no matter how severe, can ever restore life to the extinct. There is no remediation possible, no clean-up except for the bones. Extinction is a loss without limits. Extinction is a loss with no endpoint. Extinction is a deadline in the most literal sense of the word. Prevention is the only cure.

Again, if there were large numbers of unknown species hidden inside destroyed marine hotspots, the catastrophic result is magnified, and we have a "First Order Grave Error," whereas relatively low numbers of unidentified species yield a low multiplier effect and a "Second Order Grave Error." However, even if, contrary to all indications, there were *no* unknown species—no species at all remaining to be discovered—both the number and value of the species already identified are incalculably high.

Now for the good news. This comes when we invest in hotspots preservation and the (unknown and unknowable) facts ultimately vindicate our choice and show that we made the right move. A "soft benefit" happens when there is actually a high risk that whatever species exist in the hotspots will become extinct unless we act, but the tangible benefits those species offer are relatively low. This is a *soft* benefit because our actions will presumably save some species from extinction, and those species will confer intangible benefits in terms of a sense of well-being and moral satisfaction from having done the right thing. If there are many unknown species, our benefit is multiplied and we have a "First Order Soft Benefit," while the converse (few unknown species) yields a "Second Order Soft Benefit."

Where our investment in hotspots preservation finds both a high overall risk of extinction for species therein and high tangible overall value for those species, we hit the "Jackpot." Our dollars will buy the preservation of species that will pay us back manifold, the ecological equivalent of winning the lottery or hitting a jackpot on a slot machine. If there are multitudes of unidentified species in the hotspots, the tangible value of these will be multiplied further, rewarding our investment with a "First Order Jackpot," while small numbers of such species would present a "Second Order Jackpot." Either way, this is one of the greatest returns on investment we could ever imagine. Plus, the winnings keep on coming, generation upon generation, far into the most distant tomorrows. Not a bad bet.

There are two other possible consequences, each of which can spring from four different combinations of variables. If our decision is not to spend significant amounts of tax dollars on hotspots preservation, and it turns out that there is actually a low extinction threat facing the species in the hotspots, we have in effect made a "Lucky Wager." We have not squandered billions of dollars trying to save species that were not going to go extinct anyway. This is true regardless of the number of unknown species in existence within the marine hotspots or the practical value all the species in those hotspots, both identified and unidentified, hold for people and the planet. There is no need to spend money saving something that does not need to be saved.[67] As mentioned, four different ways of combining the possible values of our three variables can result in a "Lucky Wager" outcome.

Along similar lines, if we do opt to fund the proposed type of legislation to the tune of billions of dollars a year, it might again be the case that there is no great threat to the existence of whatever species inhabit the hotspots. Under these circumstances, the money we spend protecting the hotspots could be considered wasted, because we did not really need to be concerned about the extinction situation.[68] More accurately, I choose to call it "Unused Insurance," because it is somewhat akin to money we personally spend on various forms of insurance—life, health, homeowners, automobile collision—for any period in which we do not actually need to file a claim. We spend insurance money to cover ourselves for harmful, even disastrous, eventualities that might befall us. The fact that we may not suffer any misfortune that leads to a payout from our insurance policy does not mean that we were foolish to buy insurance in the first place. After all, how were we to know that we would be so lucky? Just as with "Lucky Wagers," there are four ways the variables can combine to hand us an "Unused Insurance" outcome, as you can see from the table.

If we examine the Decision Matrix and all of the ways in which the variables can be combined, we can develop a theory for optimal decision-making regarding the hotspots question. The results column holds the key. The most dramatic outcomes, of course, follow from the situation wherein the hotspots are in fact at high risk and contain species (known or unknown) with great tangible value. Where this set of circumstances is combined with a third factor that also has the highest value (i.e., large numbers of unknown species nestled within the hotspots), we find the most extreme outcomes of all.

None of the other results approach the magnitude of either a "Jackpot" or a "Grave Error." Although marine hotspots conservation could easily cost several billions of dollars each year, neither the "needless" expenditure nor the "lucky" saving of such amounts of money is on the same level of importance as a "Jackpot" or a "Grave Error." A "Jackpot" would mean incalculable benefits to people and this planet for countless years, while a "Grave Error" would spell disaster from irretrievably lost solutions to major health and environmental problems. Similarly, where "only" intangible value is available from hotspot species, saving or losing these species in numbers large or small can be a matter of considerable importance, but of a different and lower order of magnitude than a "Jackpot" or "Grave Error."

What would a rational decision-maker do? Or, put another way and using the phrase favored by many theoreticians, what would a rational utility maximizer do (the WWARUMD question)? If one accepts the premises, the decision whether to fund hotspots legislation is similar to the situation at issue in Pascal's wager.[69] We have two main options, and some unbridgeable gaps in our knowledge of crucial facts. The consequences for guessing wrong and making the wrong wager are far more momentous on one side than on the other.

First, consider the less consequential outcomes. The worst that can happen if we fund marine hotspots legislation where these is only a low extinction risk is that those billions of dollars are spent to protect species that would not have gone extinct even without our intervention. Certainly, those funds could have been spent on other things that might have yielded significant benefits, but most likely they would have been no more efficacious than any other tax dollars. This is a negative outcome, but no worse than any other governmental spending that eventually proves to be suboptimal, and we all know we have plenty of examples of that. The corollary of this is the impact of a decision to refrain from funding hotspots conservation where we find that no disasters result because there was only a low risk of extinction. We would have that money available to spend on other governmental programs or on reduction of the national debt, but again, probably no world-changing benefits would result. This is a positive outcome, but not of the earthshaking variety, literally or figuratively.

It may seem strange to dismiss either the expenditure or saving of billions of tax dollars annually as inconsequential, but relative to the most extreme results possible, that assessment is exactly right. This is because there is, in effect, no limit to the magnitude of either a "Grave Error" or a "Jackpot" result.

A Grave Error situation is the ultimate example of the "penny-wise, pound-foolish" syndrome. If we gamble that the hotspots in our oceans are not facing a major extinction threat and that the tangible value of the species within them is not high, there is a chance that we could be wrong. We would do nothing to stop the extinction of species, perhaps millions of species, that hold the keys to conquering deadly diseases (some of which may not yet exist), improving food production, reducing toxic pesticide use, and a vast array of other vital benefits. It would be difficult to place a dollar value on such losses, but many human lives could easily find their way onto the casualty list. If the twenty-first-century counterpart to penicillin were one of the lost opportunities, billions of dollars per year could not begin to measure the gravity of our error. Our decision not to fund hotspots preservation would literally be dead wrong.

In the same way, the upside potential of a decision to protect marine hotspots is essentially unbounded. If our funds block the extinction of numerous ocean species with great practical value, we could save the source of the next penicillin and prevent many other colossal benefits from disappearing. Again, if we liken hotspots conservation spending to buying insurance, this would be an insurance premium well spent indeed. No one could accurately assign a dollar value to such treasures. This "wager" on hotspots preservation, with all the variables aligned, could be the wisest choice humans have ever made with regard to themselves, not to mention the environment.

This set of options is analogous to those weighed in Pascal's wager. We basically have two choices—to fund marine hotspots preservation adequately or not.[70] There are important unknowns relevant to the issue of which option is preferable. The unknowns cannot be known, at least not without a huge amount of work over a long period of time. But we do know that a decision to protect hotspots in our planet's oceans has the possibility of paying immense, nearly infinite dividends, with only relatively minor negative consequences under the worst-case scenario. We also know that a decision not to protect those hotspots could lead to horrific, nearly infinite harm to people and this planet, but could only offer comparatively small rewards even under the best of circumstances. In this situation, the rational decision would be to protect the hotspots. This option eliminates the possibility of ruin while opening the door to limitless gain.

Of course, as I said before, there is a wide range of possible actual values for each unknown, on a constantly-evolving continuum stretching from very high to very low, but I have chosen only the extreme end-point values for ease of understanding. As I described in *Ark of the Broken Covenant*,[71] the results strongly suggest that the rational decision is to bet on the hotspots, and take meaningful steps to preserve them, even if the cost in terms of tax dollars appears to be quite high. This is so because the benefits of investing in conservation of the hotspots are phenomenally large if it turns out that there are many unknown species living therein that have a high practical value available to humankind, and a severe risk of extinction if we stay with the status quo. We could be ensuring the availability of indispensable medicines, foods, and genes, for all future generations for all time. Conversely, if we invest a great deal of money and effort in preserving the marine hotspots and in actuality there are not many species endemic to these regions, with little tangible value to us, and at minimal risk of extinction, the only downside is the "waste" of conservation dollars that might have been spent (or saved) for other projects. It is properly considered "unused insurance," conceptually no more a waste or a foolish investment than any of the (one would hope many) payments we make on our life insurance premiums during all the happy years we continue to remain alive.

Personally, I am rather pleased as every year passes without the need for anyone to file a claim under my life insurance policy. *That means I'm not dead yet!* I do not view the premiums I paid on the policy during that year to be a waste of money that could have been better spent on a high-definition television. I don't exclaim, "What a fool I was to squander my hard-earned money on that stupid life insurance stuff! I'm canceling my policy right now!" If I ever did blurt out anything along those lines, you can rest assured that I would never live to cancel my policy, and the insurance company would soon be writing a check to my wife—I mean, widow.

On the other hand, there is an unimaginable cost for failing to preserve the marine hotspots if they contain numerous species of high value at great risk of extinction. We could cost ourselves and our posterity untold advancements in medicine, therapies, genetic resources, nutrients, ecosystem services, and other areas, including perhaps a cure to a global health threat that might not materialize until centuries from now—truly a "Grave Error" of the first order. But if we sit on the sidelines and fail to invest in hotspots preservation, and we "get lucky" (few species, low value, small extinction risk), our only gain is in the form of saving the money and effort we could have spent on the hotspots. Even if this amounts to several billion dollars a year, it is a small benefit compared to the incalculably catastrophic losses we could suffer if we guess wrong in betting on the inaction option.

The Decision Matrix actually underrepresents the extent to which the rational decision is to invest in hotspots preservation. Because the Decision Matrix, in tabular form, devotes equal space to each of the sixteen possible combinations of extreme variable values, it can mislead readers into thinking that each of the sixteen outcomes is equally probable. This is most emphatically not the case. Some of these results are far more probable than others. This problem of apparent equality of disparate results is of the same type as a chart that depicts a person's chances of being fatally injured by a plummeting comet on the way home from work on any given day. There are only two possible results in such a table (survives another day, or killed by comet), and they would occupy an equal amount of tabular space on the printed page, but the probability of the former outcome is, thankfully, much higher than the likelihood of the latter tragic event.

As explained previously in this book, it is much more likely that there are numerous, even millions, of unidentified species currently living in the marine hotspots than that these hotspots are really not centers of profuse biodiversity still awaiting our discovery. It is also very probable that the extinction threat in our oceans is real and significant, given what we know about the horrific effects wrought on coral reefs and other known marine population centers by overfishing, dredging, trawling, pollution, sedimentation, and other human-made stressors.[72] Recent discoveries have revealed very high rates of endemism in small areas such as seamounts, which are extremely vulnerable to trawl damage.[73] Even in deep ocean areas, there is evidence that new technologies are making it both a possibility and a reality to exploit the previously unexploitable biodiversity in these waters via demersal fishing/trawling, to devastating effect.[74]

Only a truly Orwellian brand of doublethink could label as progress the development of fishing methods that do to the benthic habitats what modern clear-cutting has done to so many forests (on a scale 150 times as severe), but it is this "progress" that has brought mass extinction to the seas.[75] An area as large as the Gulf of Maine and the Georges Bank combined is trawled each year, 150 times the amount of forest annually lost to clear-cutting globally. Put another way, the seabed area disturbed by trawling on an annual basis is as large as the terrestrial equivalent of Brazil, India, and the Congo added together![76]

However, there is also a positive side, in light of the large numbers of marine species and habitat types that still exist, including life forms adapted to extraordinary niches such as hydrothermal vents and the abyss. That is, it would be surprising if there were not highly valuable genetic resources, natural medicines, potential sources of food, and other boons waiting to be discovered in such areas.

Therefore, the results that are linked to high, rather than low, values of each of the three Decision Matrix variables are far more probable than the converse outcomes. In terms of probabilities, it is much more likely that either a "First Order Grave Error" or "First Order Jackpot" will occur than a "Lucky Wager" or an "Unused

Insurance" result. In fact, *all* of the combinations with either two or three "High" values of the variables are significantly more probable that any of the combinations with two or three "Low" variable values. This means that the tilt in favor of betting on the hotspots is much more pronounced than is apparent from a cursory glance at the Decision Matrix. The extreme results are far likelier to fall in favor of hotspots preservation than the opposite.

If I were to depict this graphically, with a pie chart, the situation would be immediately obvious. Reflecting the factors I just mentioned, the pie would definitely not be divided into sixteen equal slices, like a bicycle wheel with regularly spaced spokes (or a fairly divided blueberry pie to serve sixteen people equally). Instead, the sixteen slices would be of quite different sizes, with some very large, others very small, and some in between. The most generous slices would represent, of course, the two outcomes with the highest probability: "high" values for all three variables. The two stingiest slices (i.e., the least likely results) would correspond to "Low" values for all three unknowns. Moreover, *all* of the larger slices would be reserved for outcomes with more than one "high" variable, constituting by far the majority of the pie, much more than half. The image of this pie with all those unequal slices should help us visualize how the variables point decidedly in one certain direction: it is *much* more likely than not that hotspots preservation is a smart move.

The Hotspots Wager and Decision Matrix tell us that the optimal choice regarding the marine hotspots is very clear. It is by far the most prudent, rational decision to invest in systematic, vigorous, and comprehensive hotspots preservation throughout the world's oceans. To do otherwise is shortsighted at best and wildly reckless at worst, a global game of Russian roulette with a very large harpoon pointed at the source of all life.

Could this paradigm actually change minds and make a difference? Where is the committed constituency pressuring Congress to remedy the Sixth Extinction? How many K Street lobbyists are wooing legislators toward earmarks for hotspots preserves? The antidote for the inertia that has so firmly mired hotspot protection in the mud of inaction is education. The hotspots concept is still very new, even within the scientific community. My books are the first within the legal community to focus entirely on hotspots, and political leaders, legislators, and members of the general public are unlikely to have had much if any exposure to the hotspots concept as yet. The evidence of the immense importance of hotspots and the threats to their continued viability is formidable, and might prove persuasive to many people if they were aware of it, particularly if they logically view all the factors in context, along the lines of my proposed variation of Pascal's wager. There is much work to be done in that regard, and there is no time to waste. No one knows when the invisible extinction clock will reach midnight for each of the many thousands of species at risk, but that hour is creeping inexorably nearer.

Yet all is not lost, and ideas do have the potential to transform history. I started this book with a quotation from Thomas Paine, uttered in 1776, about our power to begin the world over again. In that same year, Adam Smith published his famous and tremendously influential book, *An Inquiry into the Nature and Causes of the Wealth of Nations*. That pivotal book gave wings to key ideas about the prerequisites for "the necessaries and conveniencies of life"[77] that Smith believed constituted the wealth of a nation. The United States and other nations pursued Smith's ideas and shaped their political, legal, and economic systems accordingly. Now, more than two centuries later, the world must deal with some of the same challenges that existed in Smith's time, but also with some new, dramatically different, issues on a global scale. These issues affect "the necessaries and conveniencies of life" on a different level, driving to the core of life with a capital L—Life on Earth.

In our aggressive pursuit of Adam Smith's vision of economic liberty, we have been depriving much of the life in our oceans of that which is necessary for its survival, and the whole world is paying the price. Our execrable record of plundering this planet's natural wealth for immediate economic gain proves where our priorities lie. Our actions toward Earth's living heritage stretch from deliberate to delinquent and from blunder to blunderbuss as we dismantle the natural realm with staggering abandon. We are using every weapon of intellect and ignorance, both actively and passively, to kill and ruin our oceans. The War on the Water World may have been declared only by default, and those in the front ranks may insist that they are only supplying an eager market with seafood and energy, but all of the acts we commit and omit are adding up to a blitzkrieg in the benthos.

The wealth of nations means more than money, more even than human energy and talent. Today we should understand that the world's irreplaceable biodiversity is a very real form of wealth, which is at least as necessary as any other treasure for the preservation and progression of humankind. The hotspots are the crown jewels of planet Earth, and the marine hotspots in particular are the most spectacular pearls, the wealth of nations in its rarest and most precious form. And it is quite fitting that Adam Smith crafted his famous title in the plural, using the plural word "nations" rather than merely the singular form "nation," because the hotspots must be understood as belonging in some sense to all the world, all nations, all peoples. They are our mutual inheritance, and our mutual responsibility. Of all the great unknowns linked to the hotspots, the greatest of all is this: Will we take the actions necessary to save them before we lose them forever?

FIVE

The Greatest Unknown

Life in the Earth's oceans can no longer be entrusted to a yawningly porous safety net. This tattered safety net—the illusion of protection conjured up by the patchwork combination of international and national laws—is no match for the real commercial fishing nets and other threats that are all too often inescapable and indiscriminate. We can and do pretend that we've got the whole world in our hands and that all is well, but the safety net of laws we have stitched together will not hold life in our oceans, and the claim that it is good enough will not hold water. The truth is that we are not protecting life in the oceans. On the contrary, we are waging World War III, the War on the Water World, and giving it all we've got.

In this book I have shown that the oceans are home to a stunning array of life forms, including species, phyla, and even an entire kingdom adapted to some of the most extreme conditions on the planet. Marine biodiversity extends from sunlit, nearby coral reefs to the deepest, most impenetrably dark abyss, and from hyperheated hydrothermal vents to the most frigid waters. The amazing spectrum of evolutionary adaptations represented by life in these conditions is without parallel on land.

But the vastness of the oceans is both their greatest strength and their most acute weakness. It has for many centuries caused people to think of the oceans as inexhaustible resources and bottomless garbage dumps, immune to anything we do to them. This is exacerbated by the fact that so large a share of the oceans' expanse is legally international territory, not within the jurisdiction of any nation. As a global "commons," the oceans at once seem to belong to everyone and no one. We have treated them accordingly for too long.

Modern technologically sophisticated commercial fishing has inflicted tremendous damage on major portions of marine biodiversity. We have become much more effective at locating and catching the seafood species we want, using satellites, sonar, and computer-aided techniques to get to targets previously safe from our more primitive efforts. We are now more proficient at finding, chasing, and killing the seafood we want than anyone had ever dreamed of being throughout all of human history. Through the widespread and strategically directed employment of trawls, dredges, immense nylon nets, and other methods, we have also become far more effective at catching and killing huge numbers of unwanted species, resulting in appalling losses of by-catch on top of the vast take among the ones we are trying to get. The combined effect is to eviscerate large segments of the once-teeming marine food web in key regions.

Land-based activities have also caused enormous harm to vital marine habitats such as coral reefs and other parts of the continental shelf. Pollution runoff from agricultural, silvicultural, mining, industrial, and developmental activities, as well as sedimentation, have profoundly altered these sensitive ecosystems, with devastating effects on the biodiversity endemic to them. In a remarkable example of our efficiency at purging the planet of its biodiversity, we are simultaneously wiping out terrestrial hotspots in the tropical forests and catastrophically increasing the amount of runoff from those lands into the near-shore marine hotspots.

Marine pollution farther from shore has been another destructive factor. Both deliberate dumping from ships and accidental discharges, spills, and leaks have introduced large amounts of oil, organic waste, and chemicals into the oceans. Some of these are short-term dramatic incidents, and others happen little by little, day by day, to nonetheless deadly effect. Noise pollution, and the effects of climate change, add to the habitat-altering crisis.

As on land, biodiversity in the expanse and depth of the oceans is most definitely not uniformly distributed. There are areas of concentrated biodiversity, where a disproportionate number of species and higher taxa are endemic to a relatively small geographic region. These marine hotspots are epicenters of biodiversity, with incalculable significance for the planet as a whole. Yet, just as on land, the legal regime does not explicitly recognize hotspots, and in no way focuses legal protection or conservation resources on what should be high-priority areas. There is an ongoing crisis in marine biodiversity, amounting to a mass extinction of historic proportions, and the law has neither prevented nor halted it.

This is a colossal failure of the law in a matter of unimaginable importance. There are numerous international legal agreements that purport to deal with the health of the marine environment and its biodiversity to one degree or another. But because of ambiguous, loophole-ridden strictures, lax or nonexistent enforcement, and the refusal of important nations to become signatories, all of these conventions and treaties in the aggregate have been inadequate. The mass extinction, indeed, has largely begun during the last few decades when this network of laws was either already in effect or assembling the final pieces. And the individual laws of the many nations with coastlines and/or major fishing and shipping industries have mostly done very little to fill in the gaps.

I have likened this agglomeration of laws to a placebo prescribed for a patient with a serious illness. All those laws, with grandiose, encouraging names such as the Convention on Biological Diversity, the World Heritage Convention, and the United Nations Convention on the Law of the Sea, have created a very dangerous illusion that whatever problems might once have threatened our marine life have been solved. But the mass extinction rages on, and the presumptive solution is an illusion, a placebo. A placebo might temporarily help a desperately sick person feel better psychologically, but the reality and gravity of the situation will ultimately become inescapably apparent. Without real medicine that has actual power to cure a person's malady, a placebo only puts a happy face on an ugly truth. And if reliance on a placebo causes a patient to forego other therapy, it can be a deadly deception, a prelude to a death mask.

It is unrealistic to expect mere amendments to the existing international legal structure to bring about the needed sea change in marine biodiversity law. There are fundamental, systemic flaws in the current regime that cannot be remedied by tinkering and fine-tuning around the edges. There are no edges. There is no "there" there.[1]

This unsatisfactory and deceptive situation regarding our legal response to the mass-extinction crisis reminds me of a bit of dialogue from the film *Amistad*. In one scene, Cinque, an African man illegally captured and brought to the United States as a slave, angrily speaks out against the American legal system that has kept him in chains for so long. Cinque cries out, "What kind of place is this? Where you almost mean what you say? Where laws almost work? How can you live like that?"[2]

In this book I have argued for a dramatic departure from the legal status quo. More importantly, I have demonstrated why rational decision-makers should choose to adopt this proposal if they were made aware of all the relevant factors and the proper conceptual paradigm. A counterintuitive approach need not remain counterintuitive once the appropriate parameters are assembled and evaluated with due weight to the correct variables—and once we look at the problem from the right perspective.

I maintain that my Hotspots Wager and its corresponding Decision Matrix provide the analytical key to the crisis in marine biodiversity law. These tools clarify the role that uncertainty must play in determining the right course of action. This is essential, because the marine hotspots issue features huge unknowns, vast and unfathomable gaps in the relevant information base. There is so much we do not know about our planet's oceans and the life within them.[3] There remain, even in the twenty-first century, large parts of the marine world that might as well be on Neptune for the pitiful extent of our knowledge of them. The oceans are so colossal, in terms of their width, length, and depth, and so inhospitable to human penetration, that we have scarcely begun to explore them. Deep within the unconquerable, eternal darkness, how has life evolved to deal with bone-crushing water pressure, unparalleled extremes of temperature, and other intense environmental challenges? How many life forms exist in the oceans, unknown to us?[4] What benefits might these living things offer, if only we knew about them? How close to midnight are these species, as the extinction clock ticks on toward a secret deadline?

The answers to these vital questions are unknown and probably unknowable long into the future. But the Hotspots Wager and Decision Matrix take those unknowns and prove that they point to a specific answer, one that rationally takes into account all the variables and potential gains and losses inherent in our legal options. In the case of the marine hotspots, the answer is a very different one from the status quo. In the world's oceans today, we see the results of the standard autopilot legal approach: a mass extinction that threatens many of the most unusual, most important, and least understood living things in existence. We are rapidly killing our oceans, and in them numberless ocean dwellers we have never even named.

Is this deadly trend irreversible? That is the greatest unknown of all. And a big part of the reason why it is unknown is the fact that so much of the answer depends on those most unpredictable of living creatures—human beings. There is ample cause to be pessimistic, even fatalistic, given certain realities. For one, it appears that the U.S. Congress almost always needs a crisis, whether real or imagined, to galvanize it into legislative action. Congress is the antithesis of a proactive organization. Especially in the category of environmental protection, it is doggedly reactive, mired deep in inertia until and unless events forcibly blast it out of its foxhole.

Perhaps this is because members of Congress are unusually diligent and dedicated, working every available moment to discover and address in the best possible way all of the most pressing needs of the American people. Such is their single-minded devotion to the public good that it is the rare initiative indeed that will rise to the top of their mountainous stack of worthy projects. Or not.

It is also possible that members of Congress are often obsessed with their own reelection, and determine their legislative priorities largely on the basis of those causes that they believe are most apt to affect their political prospects. According to this admittedly skeptical, even cynical, point of view, Congress will not pursue legislation if there is less than overwhelming evidence of a significant public demand for it. If there is a large or highly vocal constituency behind a particular cause, Congress is much more likely to pay attention. It is a simple fact of life.

This phenomenon has been replicated many times in the field of federal environmental law. Most of the major laws were enacted—often hastily and without adequate research and deliberation—in response to the furor *du jour*. Some of these epidemics of popular outcry were rooted in legitimate dangers, while others were more the product of media spin, but in each case there was sufficient public demand (at least as perceived by a majority of both houses of Congress) to spur a successful bout of legislative activity. Even a book, Rachel Carson's *Silent Spring*, has been a significant factor in jump-starting the Congressional engine (in that case, to pass the Endangered Species Act).[5] But controversies such as the incidents involving Love Canal, Three Mile Island, and *Exxon Valdez*, plus the spectacle of fires on the Great Lakes, deadly air-pollution episodes, and medical waste washing up on popular beaches, have been key factors giving rise to the modern panoply of environmental regulation.

We may not have such statutes as the Comprehensive Environmental Response, Compensation and Liability Act[6] (often erroneously referred to as "Superfund"), the Federal Water Pollution Control Act[7] (often called the Clean Water Act), the Clean Air Act,[8] the Resource Conservation and Recovery Act,[9] the Oil Pollution Act,[10] and the Medical Waste Tracking Act[11] if it were not for the amazing power of voter agitation to prod the sleeping congressional giant out of its lethargy. But there is no such outcry against the current mass extinction. On the contrary, where there isn't outright ignorance of it, there is almost always abject apathy.

This may be partially attributable to the fact that the current mass extinction, particularly within the oceans, is virtually invisible to most people. Humans seem to be shortsighted and focused on the near-term and most immediately relevant concerns as a general rule, and these tendencies are probably more pronounced now than ever before. Modern mass-media entertainment and communications condition us to expect and even demand swift if not instant gratification, and dramatic, highly visible results from our efforts. To say the least, the contemporary marine extinction spasm fails to live up (or die up) to these preconceived expectations of what a crisis looks like. The millions of people who flock to special-effects–laden films such as *The Day After Tomorrow* have been taught that environmental calamities—even notoriously gradual ones like the onset of an Ice Age—happen with shocking and horrifying suddenness. They believe that you can actually *see* the new Ice Age sweeping rapidly down the hall, so swift and deadly that you have to rush into a room and quickly slam the door behind you before you are freeze-dried in your tracks. When people "learn" that an Ice Age is something they have to outrun to avert a sudden and immediate frozen death, how can drab, unexciting reality compete?

It is decidedly anticlimactic to tell a media-age person what a real mass extinction looks like. When I tell groups of intelligent adults, "If you want to see a mass extinction, peek out your window," I can actually see the disappointment on their faces. How can a mass extinction appear to be business as usual? How can one of the six most momentous catastrophes in our planet's multibillion-year history look so...normal? How can the extinguishment of tens of thousands of species be so...boring? Any director who would make a major motion picture out of an event this bland would without a doubt be headed for direct-to-video.

Although extinction is absolutely and eternally "lights out," the lights go out so gradually that, from the human perspective, we rarely notice any dimming at all. It is a bit like the old fable of the frog and the hot water. If you throw a frog into a pot of scalding water, it will instantly feel intense pain and waste no time in leaping out of the water to safety, redder but wiser. That one is a no-brainer, even for a frog. But if you put a frog in a pot of cool water and place it on the stove, very slowly turning up the heat, the frog will not feel any heat at first. The temperature climbs so subtly that there is never any sudden shock of sensation to set off the frog's internal alert system. Even when the water becomes as scalding hot as in the first example, the frog still will not try to escape, because the threat has crept up so imperceptibly. The hapless frog will cook to death without ever attempting to flee, lulled into fatal complacency by the gradualism of its demise.

Extinction is like that—even mass extinction. As the character Antonio Salieri chided Mozart in the play and film *Amadeus*, most people want a good loud BANG at the end of a piece of music, "so they know when to clap."[12] We expect something sudden, dramatic, and unmistakable to mark the conclusion of anything important. Especially when some law professor condescends to climb down from his Ivory Tower long enough to beg the public to buy and read his books warning of impending doom, there should be a big, obvious payoff that proves incontrovertibly that there really is a good reason to get so excited. If most species go extinct not with a BANG, but with a long, slow whisper, who will notice? If only the biodiversity hotspots would oblige and grow hotter a lot faster people might wake up and pay attention. Instead, like the complacent frog calmly soaking in an ever more dangerous bath, we cannot even tell that the hotspots are secretly reaching the boiling point. The warning signals are there, for those who know what to look for, and who have their eyes open. But for most, it is all so blissfully invisible.

But that's life. And that's death. All around us, we see deadliness disguised as steadiness. In the real world, every mass extinction— even the ones like the K-T event widely believed to have been precipitated by an enormous collision with a meteor or comet—took many thousands of years to exact its terrible toll. Life is astoundingly tenacious, and most species will hang on in last-stand mode, even as the living dead, for centuries or millennia as their critical habitat shrinks inexorably toward nothingness. During all those many hundreds and thousands of years, the numbers of members of the doomed species will rise and fall, often numerous times, until the end finally arrives. And if human beings are around to witness the extinction, scores of generations of people will rise and fall while the species lurches gradually toward oblivion. We are probably experiencing this phenomenon right now, in cases such as the tiger. The tiger may well be committed to extinction already, sad to say. But most people will never know that, and if they hear it they will refuse to believe it. They see these beautiful animals, alive and well, in zoos or circuses or Las Vegas shows, with some regularity. They can see film footage of tigers in the wild, even today. How can they be among the living dead? Impossible, scare-mongering, fanatical nonsense! Just another heaping helping of sky-is-falling false alarms from tree-hugging zealots!

Under these gloomy circumstances, what hope is there for the type of intense public outrage necessary to spark congressional action to stave off this modern mass extinction? A mass extinction does not only affect *our* nation (and not even particularly our nation), but rather has an effect that is diffused over the entire world . . . so what's in it for us to take the lead and get the lead out? Moreover, a mass extinction does not strike with a sudden and shocking thrust but with a glacierlike gradualism, an imperceptibly slow erosion of life. The natural human reaction under these circumstances is to deny that there is any problem, at least until the tipping point is reached and the evidence becomes undeniable—but by then, it will be too everlastingly late.

And what a tipping point this would be, at the culmination of the long, slow extinction death march. Thousands of marine species would disappear forever, taking with them untold promise of benefits to humankind and the planet's well-being. But even when this point is reached, and myriad species tip over extinction's cliff into oblivion, the outward appearance of it all may still be ambiguous at worst, or even transparent, to any casual onlooker. This is because so many of these species are never seen by anyone anyway; they are small, inconspicuous, drab, and live at depths and at distances from shore that virtually preclude human observation. In many cases, these species are and have always been unknown to people, and have no scientific name. So who would miss them when they are gone? We never knew they were here in the first place. At most, we might be aware of the disappearance of some familiar, euphotic-zone species, much like what we are experiencing now with the devastating drop in many larger fish populations. This can serve as a symptom of much broader and deeper extinction, akin to the proverbial canary in the mine shaft or the tip of the iceberg. The little we can see is a warning, an indicator that the problem is far more extensive than that which is readily apparent. But this situation by its very nature makes it less likely that many people will become aware of the problem, much less be provoked into demanding corrective action.

Thus, even the final act of the most cataclysmic marine extinction imaginable, pulling numberless ocean species into the black hole in a relatively short span of time, would be unseen by human eyes, taking place with insidious subtlety far beneath the water's surface. It scarcely needs to be said, but I will say it anyway: This is not the type of event that usually sets the public's outrage meter pegging into the red zone. For the great majority of citizens, it would be as if the situation were entirely normal, and there would be no pressure on Congress to do anything about it.

But even if some dedicated scientists should succeed in documenting the marine mass extinction, and getting the tragic information to a fairly large audience (perhaps with a television program or film documentary), would that suffice? I have tried to demonstrate how every person has a powerful self-interest in ensuring the survival of as many species as possible. There are potent utilitarian reasons why people should fight to preserve species that might today or someday supply important medicines, therapies, immunities, foods, ecosystem services, and genetic advancements. So, assuming that people are made aware of the marine mass extinction now ensuing, would it not be a rational decision to demand legislative action to halt the carnage? That, after all, is the purpose of my Hotspots Wager and Decision Matrix—to prove that investing in hotspots preservation is the smart course of action.

Sadly, the answer is anything but encouraging. Legal theorists, economists, social scientists, and other highly educated experts are fond of the concept of people as rational utility maximizers. That is, they formulate their theories and policy recommendations based on the idea that most people will intelligently and logically select whatever options are best for themselves in any situation, choosing the most favorable course of action for the furtherance of their own self-interest. Thus, people will respond rationally and predictably to any set of incentives, disincentives, threats, risks, rewards, and inducements, using logic and reason to weigh the probability and the magnitude of each possible outcome in determining the optimal decision for themselves. But, as that great philosopher Jiminy Cricket put it (in a slightly different context) in the classic film *Pinocchio,* "A very lovely thought, but not at all practical."[13]

Why should we delude ourselves that people, individually or collectively, will behave as rational utility maximizers with regard to marine hotspots preservation? Let us depart from the clean-room artificiality of the academician's imaginary universe for a moment and step into the harsh light of the real world for a bit of reality therapy. How many people do you know, in your own circle of acquaintances, who, in their own lives, regularly make choices that virtually guarantee to destroy their health, economic well-being, families, and personal happiness? With reference to major life decisions of undeniable and obvious importance to the individual, how often do you see people making blatantly stupid choices on matters of mate selection, use of illegal drugs, excessive consumption of food, overuse of alcoholic beverages, smoking, irresponsible gambling, avoidance of exercise, unprotected casual sex, violation of the law, and many other such opportunities to harm themselves and their loved ones? I would bet (but not irresponsibly) that you, gentle reader, know several people who routinely—not occasionally, but routinely—make colossal blunders in one or more of these vital categories of decision making.

It seems that each of us is afflicted with one or more "blind spots" in which we repeatedly make idiot-level choices. Some of us have more blind spots than others, but we all have at least one. It is appalling to learn how people who may be exceedingly intelligent, accomplished, respected, responsible, successful, and well educated will still have that Achilles' heel that leaves them vulnerable to achingly obvious dangers. There are so many examples, and the phenomenon is so ubiquitous, that there is hardly any need to provide examples. Even if we confine ourselves to the actions of recent presidents of the United States, there will be no shortage of exemplars. The hard truth is that no amount of natural intelligence, experience, education, and upbringing can immunize a human being from the innate right to be wrong—even howlingly, amazingly, consistently wrong—in one or more compartments of life's multifaceted aspects.

It is full disclosure time. I myself am not immune to this defect. I am indebted to my wife, Marcia, for identifying one particularly notable blind spot that afflicts me: my tendency to devour generous quantities of high-calorie, high-cholesterol, high-grease cuisine. Marcia has declared that I eat in such a fashion so as to maximize the probability that my obituary will feature words such as "massive" and "suddenly." Yet in other areas of my life, I am quite rational (or so I like to believe, anyway).

As my reference to our presidents suggests, the penchant for ir-rational, even self-destructive decision-making is part of the human condition. It extends from the humblest individual all the way up to the very pinnacle of human achievement. Not everyone has the same Achilles' heel, of course. For some it is an appetite for sexual ad-venture, for others a craving for the sensations delivered by cigarettes, drugs, or liquor, and for still others the hunger for the thrills and dangers posed by risky sports or excessive gambling. You can see evidence of this in the life experiences of those whom you personally know, as well as in news accounts of the lives of the rich, powerful, and famous. No amount of letters after a person's name, no military grade, no political title, no number of years, and no quantum of wealth suffices to confer upon any of us one "get out of jail free" card, or one "immunity to unwise decisions" card either. So how realistic is it to suppose that exposure of, say, all members of Congress to my hotspots Decision Matrix will inspire them to take the appropriate steps to stem the trend toward mass extinction?

It does not help that most members of Congress—and most presidents—have little or no scientific background. For politicians who do not know a phylum from a phallus, and think that kingdoms are only for politicians in other countries, can we reasonably expect an appreciation of the higher-taxa diversity of the oceans, let alone a determination to save it? For half-educated political junkies who have no conception of what it means to have so many classes and phyla—and even an entire new kingdom, the Archaea—endemic solely to the marine realm, can the world realistically hope for enlightened action?

One of the most intractable obstacles to halting our self-inflicted mass extinction is the inestimable loss of "common knowledge" among today's people. For the last few decades we have taught less and less, and have expected and demanded less and less from our students at all educational levels. Self-esteem, thorough indoctrination in the quasi-religious dogma of nonjudgmentalism, and the ferocious pursuit of untrammeled personal liberty are about all we really teach anymore. Today's pupils, even at the college and graduate-school level, and even at prestigious institutions of what was once called higher learning, know ever less about ever more. Even as information overload has swamped us with omnipresent access to torrents of communication as we careen down the information superhighway, we actually learn very few facts or rigorous principles along the way, opting instead for the intellectual equivalent of fast food snatched from the drive-through window. Most of us never even bother to step out of our rented car.

We now know shockingly little of history, culture, mathematics, or science. Not only do we not know the facts, we do not know how to think, how to reason with precision of thought. And without that once-universal foundation, we are utterly unequipped to comprehend the magnitude and meaning of the crisis in biodiversity we have so thoughtlessly spawned. The one thing we have *not* forgotten is how to kill.

It is always easier to destroy than to create. It is also quicker. Throughout human history, barbarians of diverse races, creeds, and political allegiances have brutally and brutishly demolished irreplaceable cultural treasures that took many years for the highest and best of civilized people to craft. The senseless acts of humans desecrating the pinnacles of human achievement are not, sadly, only tragic chapters from distant history. From the torching of the Library of Alexandria, to the defacing of the Sphinx, to the obliteration of gigantic ancient Buddhist statues[14] of Bamiyan—it is all just different lines on the same page of our long march through time. The more years of dedicated work it required, and the greater the genius it demanded, to bring these wonders into existence, the more eager other people have always been to burn and bash it all into a heap of waste, in a single spasm of hatred. Who can now restore the ashes of Alexandria's legacy, or turn back time to the day before the Sphinx was shattered? Such despoliation is as permanent as it is poisonous to civilization.

It is the same with our vandalism of Earth's living treasures, but even more extreme, in a way. Every species consumed millions of years to arrive at the present day as it now appears. No humans made them, nor could they—not even a Leonardo da Vinci in all his brilliance. Many of these species were here countless years before people even stepped onto the stage. But what required Nature thousands of millennia to create is now being ruined in only a few decades of human barbarity. Again, it is so much faster and so much easier to kill than to create. And if we cannot undo the effects of bombs and fire-brands on our human heritage, how can we begin to hope to resurrect the species we are chasing into extinction? The destruction of such magnificent, sublime masterpieces of Nature is beyond remedy once it goes too far. Extinction is a one-way, nonrefundable, express ticket to oblivion. There is no taking it back after it happens, no wiping the slate clean after species have been wiped out.

Our vandalism of life is no less reprehensible than our dismantling of civilization's gems. True, this ecological vandalism is generally motivated by greed and is incidental to the pursuit of other goals, not driven by wanton hatred, jealousy, vengeance, and prejudice. The effects, though, are every bit as everlasting, and perhaps even more costly to our posterity. The Romans of old practiced an official form of vandalism known as *damnatio memoriae*, "damnation of the memory," when later emperors would smash the statues of their predecessors and remove their names from public view. The intent was to shape the future by removing the past. This execrable policy cost us more than we can ever know of our cultural roots. But, in our ignorance and selfishness, we are no better today. Every day, we are perpetrating our own system of *damnatio memoriae* on a vast scale as we lay waste to our unique and irreplaceable living roots in the oceans. And in all too many cases, by killing species we have never even discovered, we are damning memories before we ever have the chance to form them.

It is strange and paradoxical how perishable, how fragile are some of our oldest legacies—whether cultural or natural. The crumbling, fading faces of Leonardo da Vinci's *The Last Supper* still look down from their old wall, and, however faintly, they do sometimes seem to me to be reacting in frozen horror to the death of life all around them. It is easier for most of us to understand the brittleness and delicacy of our greatest art from antiquity because we can see for ourselves the irrefutable scars of decay—on the pyramids, on the Parthenon, on fragments of statues from civilizations long ago vanished. The evidence of their tenuous and precarious continued existence is so plain that any child can know it from a moment's glance. And, in knowing the dangling, fraying lifeline that holds such art to us, we instinctively realize the importance of protecting and preserving that art for all time. But we cannot see that so much of life on Earth is equally vulnerable, although infinitely more ancient than our most venerable artworks. The evidence is there, but it is far more treacherously sly than the visible signs of *The Last Supper*'s peril. If it were obvious, if it were clear for anyone to see, we might be as eager to rescue our dying oceans as our decaying masterworks. Ultimately, both Nature and art are as easily lost as they are impossible to replace, both life itself and that which makes life worth living.

This modern mass extinction is so far beyond the stretching reach of our mundane experience that, to us, it may have the feel of an ancient myth—larger than life, unreal, impossibly exaggerated, and past belief. It is almost easier to accept as genuine the tragic and heroic deeds of Heracles and Achilles than to acknowledge what is happening to the world under the waves right now. Skeptics even deride the idea of a twenty-first-century mass extinction as an eco-myth, a scary fairy tale fantasized by fanatics. True, the magnitude of this disaster is on a scale our antiquarian ancestors would have thought only possible of gods and demigods. Who but an immortal could inadvertently reap down thousands of living things with a sword as invisible as it is unsparing? Who but a god from Mount Olympus could unknowingly demand the sacrifice of legions of species as part of a rite of homage, a forced tribute to selfish egocentric whims upon the altar of the god of money?

Only the same species that conceived of and believed in Zeus, Athena, Poseidon, and Apollo could exalt itself to such Olympian powers of life and death over innumerable other species. Only the one species with the godlike capacity to create immortal ideas could so boldly and blindly invoke the mortality of its fellow living creatures. It is as if we humans have been cruelly cursed with a tragic flaw simultaneously with our being blessed with awesome strengths. Like Achilles, we have our own vulnerability that we carry with us always, just as much a part of ourselves as our incredible ability. Homer, in his *Iliad*, could not have written a more ironic and heartbreaking tale about any of the mortals, gods, and demigods who decided the course of the Trojan War and all its heroes. But are we truly helpless pawns in a cosmic chess match played by the gods? If we have our Achilles' heel, is it not a condition of our own making that we have the power to correct, if only we had the will? Or is our curse permanent, an indelible mark on all of humanity?

If the immortals themselves are trapped in this web of blindness, what chance do the rest of us have to see what is happening and to do what it takes to stop it? Caught up in this modern myth, even the noblest, strongest, bravest, cleverest, and most heroic people—our counterparts to Hector and Odysseus—seem fated to act out the tragic roles written for them by hidden forces. We live out our lives doing what we can for ourselves, our families, our communities, our nations, and our world, but all the while the awful and secret monster we have unleashed is devouring the heart of life on Earth. Silently, under cover of darkness, and stealthily, this self-spawned suicide serpent is taking away our future while we work, play, sleep, and celebrate, oblivious to the horrible reality to which we have been blinded, or to which we have blinded ourselves.

Yet our fates, and the fate of our oceans, are not scripted for us by forces beyond our control. As with the heroes of ancient myth, we have great power for good if we will only use it wisely. This book is meant to be a contribution to that cause—just a drop in the bucket perhaps, but what is an ocean but a multitude of buckets of drops, each with a role? No one should feel insignificant, or incapable of contributing something to an eventual sea change in momentous matters, even if drop by drop. I am reminded of the concluding lines from the Lerner and Loewe musical play and motion picture *Camelot*. The words reflect hope for a better future embodied by any single child who is "One of what we all are ... Less than a drop in the great blue motion of the sunlit sea. But it seems that some of the drops sparkle ... Some of them do sparkle!"[15] After all the harm we have inflicted on Earth's oceans, hope still remains if enough of us combine our sparkling light to shine a beacon through the night.

NOTES

PREFACE

1. Thomas Paine, *Common sense*, in COLLECTED WRITINGS (Library of America, 1995).

2. John Charles Kunich, ARK OF THE BROKEN COVENANT: PROTECTING THE WORLD'S BIODIVERSITY HOTSPOTS, 6–13 (Praeger, 2003).

3. John Charles Kunich, *Preserving the womb of the unknown species with hotspots legislation*, 52 HAST. L. J. 1149 (2001).

4. John Charles Kunich, *Fiddling around while the hotspots burn out*, 14 GEO. INT'L. ENVT'L. L. REV. 179 (2001).

5. John Charles Kunich, *World heritage in danger in the hotspots*, 78 IND. L. J. 619 (2003).

6. John Charles Kunich, *Losing Nemo: The mass extinction now threatening the world's ocean hotspots*, 30 COLUMBIA JOURNAL OF ENVIRON-MENTAL LAW 1–133 (2005).

CHAPTER ONE

1. *See, e.g.,* Eric Buffetaut, *The relevance of past mass extinctions to an understanding of current and future extinction processes*, 82 PALAEOGEO-GRAPHY, PALAEOCLIMATOLOGY, PALAEOECOLOGY 169, 171

(1990); *see also* M.J. Benton, *Diversification and extinction in the history of life*, 268 SCIENCE 52 (April 7, 1995).

2. *See* Edward O. Wilson, THE DIVERSITY OF LIFE, 29 (Belknap, Harvard, 1992).

3. *See* Douglas H. Erwin, *The end and the beginning: Recoveries from mass extinctions*, TREE vol. 13, no. 9, 344, 347 (Table 1) (September 1998). Of course, the fossil record does not permit us to pinpoint the duration of such distant events, and there is considerable imprecision in these estimates.

4. Buffetaut, *supra* note 1, at 171. *See* J.J. Sepkoski, *Phanerozoic overview of mass extinctions*, 277–95, in PATTERNS AND PROCESSES IN THE HISTORY OF LIFE (D.M. Raup and D. Jablonski, eds., Springer-Verlag, 1986) for a discussion of many of the noted extinction spasms in addition to the big five.

5. *See* Juliet Eilperin, *Wave of marine species extinctions feared*, WASHINGTON POST, A1 (August 24, 2005) (summarizes the strong evidence that a marine mass extinction is now underway, yet unnoticed by most people because it is a "slow-motion disaster," both "silent and invisible").

6. *See* James T. Carlton, et al., *Historical extinctions in the sea*, 30 ANNU. REV. ECOL. SYST. 515–16 (1999).

7. *See, e.g.,* Wilson, *supra* note 2, at 243–80; Niles Eldredge, *Cretaceous Meteor Showers, the Human Ecological "Niche," and the Sixth Extinction*, 1–15, in EXTINCTIONS IN NEAR TIME: CAUSES, CONTEXTS, AND CONSE-QUENCES (Ross D.E. MacPhee, ed., Kluwer, 1999); Christopher Humphries, Paul Williams, and Richard Vane-Wright, *Measuring biodiversity value for conservation*, 26 ANNU. REV. ECOL. SYST. 93, 94 (1995); Paul Ehrlich, *Extinction: What is happening now and what needs to be done*, 157, in DYNAMICS OF EXTINCTION (David Elliott, ed., Wiley-Interscience, 1986); Stuart L. Pimm and Thomas M. Brooks, *The Sixth Extinction: How large, how soon, and where?* in NATURE AND HUMAN SOCIETY: THE QUEST FOR A SUSTAINABLE WORLD, 46–62 (Peter Raven, ed., NRC, 1997); Gary Strieker, *Scientists agree world faces mass extinction*, CNN (August 23, 2002), available at: <http://archives.cnn.com/2002/TECH/science/08/23/green .century.mass.extinction/index.html>; Michael J. Novacek and Elsa E. Cleland, *The current biodiversity extinction event: Scenarios for mitigation and recovery*, PNAS vol. 98, no. 10, 5466–70 (May 8, 2001), available at: <http:// www.pnas.org/cgi/reprint/98/10/5466>.

8. *See generally* Norman Myers, THE SINKING ARK: A NEW LOOK AT THE PROBLEM OF DISAPPEARING SPECIES (Reader's Digest Young Families, 1979); *see also* THE GLOBAL 2000 REPORT TO THE

PRESIDENT: ENTERING THE TWENTY-FIRST CENTURY, 37 (Pergamon Press, 1980), which projected the extinction of between 0.5 and 2 million species (considered by the authors to amount to 15 to 20 percent of all species on Earth) by the year 2000, mostly as a result of habitat destruction, but also in part because of pollution. This mass extinction was described as without precedent in human history. The authors hypothesized that insects, other invertebrates, and plant species, many of which are unclassified and unexamined by scientists, would bear the brunt of the losses.

9. Norman Myers, *The biodiversity outlook: Endangered species and endangered ideas*, xxviii–xxix, in SOCIAL ORDER AND ENDANGERED SPECIES PRESERVATION (J.F. Shogren and J. Tschirart, eds., Cambridge University Press, 2001). Myers argues that we are already in the midst of a major mass extinction, even by conservative estimates.

10. David Jablonski, *Mass extinctions: New answers, new questions*, 43–61, in THE LAST EXTINCTION (Les Kaufman, et al., ed., MIT Press, 1986).

11. Ehrlich, *supra* note 7, at 158–59.

12. Occasionally, previously unknown species of mammals or birds are discovered even today. For example, during the 1960s a small population of an undescribed species of cat was found on the island of Iriomote, near Okinawa, Japan (*Id.*). Four species of mammals have recently been discovered in the remote Annamite Mountains along the border between Vietnam and Laos, including a large, cowlike animal called a saola or spindlehorn (Edward O. Wilson, *Vanishing before our eyes*, TIME, 29–30 [April–May 2000]). Generally, though, the size and diurnal lifestyle of most mammals and birds makes it less likely that they can exist without being detected by humans.

13. Nigel E. Stork, *The magnitude of global biodiversity and its decline*, 7, in THE LIVING PLANET IN CRISIS: BIODIVERSITY SCIENCE AND POLICY (Joel Cracraft and Francesca T. Grifo, eds., Columbia University Press, 1999) (hereinafter LIVING PLANET).

14. Ehrlich, *supra* note 7, at 158–59.

15. Wilson, *supra* note 12, at 34.

16. *Id. See* Edward O. Wilson, *The current state of biological diversity*, 3–18 BIODIVERSITY (1988).

17. For some recent estimates, *see, e.g.*, Robert M. May, *The dimensions of life on earth*, 30–45, in NATURE AND HUMAN SOCIETY: THE QUEST FOR A SUSTAINABLE WORLD (Peter Raven, ed., NRC, 1997) (estimates 7 million species worldwide, with a range from 5 to 15 million plausible); Stork, *supra* note 13, at 10–21 (employs various factors, taxon by taxon, in arriving at a rough estimate of 13.4 million species); Paul Williams,

Kevin Gaston, and Chris Humphries, *Mapping biodiversity value worldwide: Combining higher-taxon richness from different groups*, 264 PROCEEDINGS OF THE ROYAL SOCIETY, BIOLOGICAL SCIENCES 141–48 (1997) (credits an estimate of 13.5 million species); Humphries, et al., *supra* note 7, at 94–95 (accepts a range of 5 to 15 million species); T. Erwin, *Tropical forests: Their richness in Coleoptera and other arthropod species*, 36 COLEOPT. BULL. 74–75 (1982); Benton, *supra* note 1; P.M. Hammond, *The current magnitude of biodiversity*, 113–28, in GLOBAL BIODIVERSITY ASSESSMENT (V.H. Heywood, ed., Cambridge, 1995) (estimates 12 million species total); Norman Myers, *Questions of mass extinction*, 2 BIODIVERS. & CONSERV. 2–17 (1993) (mentions several estimates and concludes that we can be fairly certain that there are at least 10 million species today).

18. *Id.*

19. *Id. See* V.H. Heywood, et al., *Uncertainties in extinction rates*, 368 NATURE 105 (1994).

20. *See* Jonathan E.M. Baillie, Craig Hilton-Taylor, and Simon N. Stuart, eds., *2004 IUCN red list of threatened species: A global species assessment*, 34, available at: <http://www.iucn.org/themes/ssc/red_list_2004/GSA_book/Red_List_2004_book.pdf>.

21. *See* Norman Myers, ed., GAIA: AN ATLAS OF PLANET MANAGEMENT, 64–93 (Doubleday, 1993).

22. These zones will be recognized by some particularly cultured readers as mentioned in a song performed by the schoolteacher, Mr. Ray, in the phenomenally popular animated film *Finding Nemo* (Disney/Pixar, 2003), from which the title of chapter three of this book draws its inspiration. In the film, a young clownfish named Nemo is captured from the ocean by humans, and his father goes to heroic lengths to find and retrieve his son.

23. David A. Ross, OPPORTUNITIES AND USES OF THE OCEAN, 6 (Springer-Verlag, 1978).

24. James C.F. Wang, HANDBOOK ON OCEAN POLITICS AND LAW, 58 (Greenwood, 1992).

25. John Charles Kunich, ARK OF THE BROKEN COVENANT: PROTECTING THE WORLD'S BIODIVERSITY HOTSPOTS, 7–9 (Praeger, 2003). *See also* Hugh P. Possingham and Kerrie A. Wilson, *Turning up the heat on hotspots,* 436 NATURE 919–20 (August 18, 2005), available at: <http://www.scidev.net/pdffiles/nature/436919a.pdf> (discusses the need to use multiple criteria in rigorously determining which areas qualify as hotspots).

26. Boyce Thorne-Miller, THE LIVING OCEAN, 2nd ED, xiv (Island Press, 1999) (hereinafter LIVING OCEAN). *See also* Office of Naval Research, *Habitats: Hydrothermal vent—hydrothermal vent life*, available at: <www.onr .navy.mil/focus/ocean/habitats/vents2.htm> (describes the 1977 discovery of the first hydrothermal vent off the Galápagos Islands). For some background on the Archaea, and the diversity of scientific viewpoints regarding how they should be classified, *see, e.g.*, Ben Waggoner and B.R. Speer, *Introduction to the archaea: Life's extremists*, available at: <http://www.ucmp .berkeley.edu/archaea/archaea.html>; Neil Saunders, *Archaea Web*, available at: <http://www.archaea.unsw.edu.au/>; Kenneth Todar, *Biological identity of procaryotes* (2001), available at: <http://www.bact.wisc.edu/Bact303/The Procaryotes>; Wikiverse, *Archaea*, available at: <http://www.wikiverse .org/archaea>.

27. LIVING OCEAN, *supra* note 26, at 48.

28. *See* GESAMP (IMO/FAO/UNESCO/WMO/IAEA/UN/UNEP, Joint Group of Experts on the Scientific Aspects of Marine Environmental Protection), and Advisory Committee on Protection of the Sea, *Marine biodiversity: Patterns, threats, and conservation needs*, REP. STUD. GESAMP no. 62, 3 (1997), available at <http://gesamp.imo.org/no62/index.htm>.

29. *Id. See also* LIVING OCEAN, *supra* note 26, at 48.

30. *See* Mark Williamson, *Marine Biodiversity in its Global Context*, 3–6, 13, in MARINE BIODIVERSITY: PATTERNS AND PROCESSES (Rupert F.G. Ormond, John D. Gage, and Martin V. Angel, eds., Cambridge University Press, 1997) (details the wide variations in species estimations).

31. *Id.* According to one accepted taxonomic system, there are thirty-three animal phyla, of which thirty-two are found at least in part in the marine environment (twenty-one of them exclusively) compared to twelve phyla found at least partially in the terrestrial environment. Another taxonomic scheme recognizes thirty-five marine phyla, fourteen of which are endemic to the oceans, and fourteen phyla at least partially terrestrial, only one of which is endemic to dry land. Under this system, of the thirty-five marine phyla, only eleven are represented in the pelagic realm, with the remainder being benthic. *See* GESAMP, *supra* note 28, at 3. The phylum, of course, is generally accepted as the second-highest taxonomic category of living things, just beneath the kingdom level. Therefore, when as many as twenty-one entire phyla are found solely in the oceans, that amounts to a colossal share of Earth's biodiversity endemic to the marine world.

32. *See* G. Carleton Ray, *Conservation of Coastal-Marine Biological Diversity*, 225–26, in BIODIVERSITY, SCIENCE AND DEVELOPMENT:

TOWARDS A NEW PARTNERSHIP (F. Di Castri and T. Younes, eds., CABI Publishing, 1996).

33. *Id.* at 226. *See also* G. Carleton Ray, *Coastal-zone biodiversity patterns*, 41 BIOSCIENCE 490, 492–93 (1991) (notes that fish are by far the most diverse vertebrates at all taxonomic levels, with 3 classes, 50 orders, 445 families, and about 22,000 species now known to be in existence, with 13,200 of those species being marine).

34. *Id.*

35. In my younger days I earned a Master of Science degree in entomology. Many years later I found myself a law professor and managed to unite these widely divergent strands of my professional career by cowriting a book on forensic entomology with my graduate school mentor and advisor, Dr. Bernie Greenberg. *See* Bernard Greenberg and John Charles Kunich, ENTOMOLOGY AND THE LAW: FLIES AS FORENSIC INDICATORS (Cambridge University Press, 2002).

36. Ronald K. O'Dor, *The unknown ocean: The baseline report of the census of marine life research program*, 6 (October 2003), available at: <http://www.coml.org/baseline/Baseline_Report_101603.pdf> (hereinafter *Unknown ocean*).

37. *Id.* at 12.

38. *Id.* at 14, Figure 9 (graphically illustrates the extreme imbalance in known marine phyletic diversity compared with terrestrial counterparts).

39. *Id.* at 6.

40. *Id.* at 25.

41. *See* Claudia E. Mills and James T. Carlton, *Rationale for a System of International Reserves for the Open Ocean*, 12 CONSERVATION BIOLOGY 244, 245 (1998), available at: <http://faculty.washington.edu/cemills/ConsBiol1998.pdf>.

42. *See* LIVING OCEAN, *supra* note 26, at 56.

43. *Id.*

44. *See* Mills, *supra* note 41, at 245.

45. *See* Williamson, *supra* note 30, at 4.

46. J.F. Grassle and N.J. Maciolek, *Deep-sea species richness: Regional and local diversity estimates from quantitative bottom samples*, 139 AMERICAN NATURALIST 313–41 (1993). Grassle and Maciolek's estimate was extrapolated from a study conducted at a depth of 4,900–6,900 feet in which 226 square feet were sampled. The results showed 798 species, representing 171 families and 14 phyla.

47. R.M. May, *Bottoms up for the oceans*, 357 NATURE 278–79 (1993).

48. J.D. Gage, *High benthic species diversity in deep sea sediments: The importance of hydrodynamics*, 150–51, in MARINE BIODIVERSITY: PATTERNS AND PROCESSES (Rupert F.G. Ormond, John D. Gage, and Martin V. Angel, eds., Cambridge University Press, 1997).

49. *Id.*

50. *See* Susan Gubbay, *The offshore directory, review of a selection of habitats, communities and species of the North-east Atlantic*, 34–35, report for World Wildlife Fund (October 2002) (hereinafter *Offshore directory*).

51. *Id.*

52. *Id.* at 35–36.

53. *See Hydrothermal vent interactives—vent basics*, available at: <http://www.divediscover.whoi.edu/vents/vent-world.html>.

54. *See* University of Delaware Graduate School of Marine Studies, *Voyage to the deep*, available at: <http://www.ocean.udel.edu/deepsea/level-2/geology/vents.html>.

55. *Offshore directory*, *supra* note 50, at 35. The life span of an individual vent is generally no longer than one hundred years.

56. *See id.* at 104.

57. *Offshore directory*, *supra* note 50, at 39.

58. Paul S. Sochaczewski and Jay Hyvarinen, *Down deep; macro- and micro-flora and fauna of the deep sea thermal vents,* EARTH ACTION NETWORK vol. 7, no. 4, 15 (July 1996).

59. *Offshore directory*, *supra* note 50, at 39.

60. *See* Cindy Lee Van Dover, THE ECOLOGY OF DEEP-SEA HYDROTHERMAL VENTS, 313 (Princeton University Press, 2000).

61. *Id.*

62. *See* Lyle Glowka, *The deepest of ironies: Genetic resources, marine scientific research, and the area*, 12 OCEAN Y.B. 154, 160 (1996).

63. *See, e.g.*, Jack D. Farmer, *Hydrothermal systems: Doorways to early biosphere evolution*, GEOLOGICAL SOCIETY OF AMERICA, GSA TODAY vol. 10, no. 7 (July 2000), available at: <http://www.geosociety.org/pubs/gsatoday/gsat0007.htm>.

64. Sochaczewski and Hyvarinen, *supra* note 58.

65. *Id.*

66. *Id.*

67. Mbari, *Submarine volcanism*, available at: <http://www.mbari.org/volcanism/Seamounts/SeamountsResearchTop.htm>; *see Unknown ocean*, *supra* note 36, at 7 (Figure 4) (illustrates the various marine zones and seamounts).

68. *Offshore directory, supra* note 50, at 21.

69. *See Seamount*, Wikipedia, available at: <www.wikipedia.org/wiki/Seamount>.

70. *Id.*

71. *Id.*

72. *See* NPACI and SDSC Online, SEAMOUNTS: WINDOW ON OCEAN BIODIVERSITY vol. 5, no. 15 (July 25, 2001), available at: <www.npaci.edu/online/v5.15/seamounts.html>. *See also* National Ocean and Atmospheric Association, *Ocean explorer*, available at: <http://ocean explorer.noaa.gov/explorations/02davidson/background/missionplan/plan .html> (describes the planned mission to document and study the species located on and around the Davidson Seamount, located just 74 miles southwest of Monterey, California. The seamount is an inactive volcano roughly as tall as the Sierra Mountains (7,500 feet) and as wide as Monterey Bay (25 miles), with its peak rising to 4,265 feet below the ocean's surface).

73. *See* R.R. Wilson and R.S. Kaufmann, *Seamount biota and biogeography*, 319–34, in 43 SEAMOUNTS ISLANDS AND ATOLLS. GEOPHYSICAL MONOGRAPH (B.H. Keating, P. Fryer, R. Batiza, and G. Boelhert, eds., American Geophysical Union, 1987). *See also* Sustainable Development International, *Unique seamount species threatened by deepwater trawlers*, available at: <http://www.marine.csiro.au/LeafletsFolder/42sea mount/42.html> (notes that a recent sampling of fewer than 25 seamounts in the Tasman and Coral Sea region uncovered more than 850 species, 42 percent more than previously reported from all studies of seamounts in the past 125 years).

74. Examples include orange roughy, sharks, tuna, and swordfish. *See Offshore directory, supra* note 50, at 19.

75. *See generally* Wilson and Kaufmann, *supra* note 73, at 319–34.

76. *See Offshore directory, supra* note 50, at 18.

77. *See* WWF/IUCN, *The status of natural resources on the high seas*, 2–3 (2001), available at: <http://www.ngo.grida.no/wwfneap/Publication/Sub missions/OSPAR2001/WWF_OSPAR01_HighSeasReport.pdf>. In addition to hydrothermal vents and seamounts, other vital and often threatened deep-ocean marine areas include deep-sea trenches, deep-sea coral reefs, polymetallic nodules, cold seeps, pockmarks, gas hydrates, and submarine canyons. WWF/IUCN explains that each of these has its own geological and biological characteristics, threats, and typical locations/depths (12–13).

78. *See* Boris Worm, Heike K. Lotze, and Ransom A. Myers, *Predator diversity hotspots in the blue ocean*, PNAS vol. 100, no. 17, 9884, 9887 (August 19, 2003), available at: <http://www.pnas.org/cgi/content/abstract/100/17/9884>.

79. *See* Steve Connor, *Protect the "Serengetis of the sea" before it's too late, say biologists* (August 6, 2003), available at: <http://www.vegsoc.org.au/forum_messages.asp?Thread_ID=508&Topic_ID=10>.

80. *Id.*

81. *See* Worm, et al., *supra* note 78, at 9887.

82. *See* Callum M. Roberts, et al., *Marine biodiversity hotspots and conservation priorities for tropical reefs*, 295 SCIENCE 1280 (2002).

83. *See* Joseph H. Connell, *Diversity in tropical rainforests and coral reefs*, 199 SCIENCE 1302–10 (1978); Oceana Report, *Deep sea corals: Out of sight, but no longer out of mind*, 1–2, 7–9, available at: <http://www.savecorals.com/news/oceana_coral_report.pdf> (hereinafter *Deep sea corals*).

84. *See* John Charles Kunich, *Mother Frankenstein, Doctor Nature, and the environmental law of genetic engineering*, 74 SO. CAL. L. REV. 807, 813–22 (2001).

85. *See generally* Horst Korn, Susanne Friedrich, and Ute Felt, DEEP SEA GENETIC RESOURCES IN THE CONTEXT OF THE CONVENTION ON BIOLOGICAL DIVERSITY AND THE UNITED NATIONS CONVENTION ON THE LAW OF THE SEA (BfN, Federal Agency for Nature Conservation, Bonn, Germany, 2003), available at: <http://www.bfn.de/09/skript79.pdf> (details many notable features, including unique genetic adaptations, of deep sea habitats, particularly hydrothermal vents).

86. *See* GESAMP (IMO/FAO/UNESCO/WMO/IAEA/UN/UNEP, Joint Group of Experts on the Scientific Aspects of Marine Environmental Protection), and Advisory Committee on Protection of the Sea, *A sea of troubles*, REP. STUD. GESAMP no. 70, 4 (2001), available at: <http://gesamp.imo.org/no70/index.htm> (hereinafter *Sea of troubles*).

87. Kunich, *supra* note 25, at 36–39. For a list of the Global 200 Ecoregions, *see* WWF, *List of ecoregions*, available at: <http://www.panda.org/about_wwf/where_we_work/ecoregions/ecoregion_list/index.cfm>.

88. *See* James T. Carlton, et al., *Historical extinctions in the sea*, 30 ANNU. REV. ECOL. SYST. 515, 530–31 (1999).

89. *See* Mills and Carlton, *supra* note 41, at 245.

90. *See* Carlton, et al., *supra* note 88, at 532. Carlton, et al., explain that although the world's oceans are immense, specialized marine environments are much smaller. Coral reefs (considered the marine equivalent of tropical

forests because of their profuse biodiversity) occupy only about 231,600 square miles (0.1 percent of Earth's surface), a small fraction of that covered by terrestrial tropical forests.

91. *Id.* at 529–30. Restricted geographic distribution, limited habitat, and poor dispersal abilities are major factors (as on land) that render marine life forms prone to extinction.

92. *See generally Sea of troubles, supra* note 86, at 3.

93. *See* Carlton, et al., *supra* note 88, at 529–30.

94. *Sea of troubles, supra* note 86, at 3. *See also* GESAMP, *supra* note 28, at 8–13.

95. *See generally* Ratana Chuenpagdee, et al., *Shifting gears: Assessing collateral impacts of fishing methods in U.S. waters*, FRONT ECOL. ENVIRON. vol. 1, no. 10, 517–24 (2003), available at: <http://www.mcbi.org/Shift ingGears/Chuenpagdee_et_al_(Frontiers).pdf>.

96. *See* J.A. Koslow, et al., *Continental slope and deep-sea fisheries: Implications for a fragile ecosystem*, 57 ICES JOURNAL OF MARINE SCIENCE 548, 549 (2000), available at: <http://communications.fullerton.edu/forensics/documents/fisheries%20-%20koslow.pdf>.

97. *Id. See also* Steven A. Murawski, *Definitions of overfishing from an ecosystem perspective*, 57 ICES JOURNAL OF MARINE SCIENCE 649, 652–53 (2000), available at: <http://communications.fullerton.edu/forensics/documents/fisheries%20-%20murawski.pdf>.

98. *See* Paul K. Dayton, et al., ECOLOGICAL EFFECTS OF FISHING IN MARINE ECOSYSTEMS OF THE UNITED STATES, 8 (PEW OCEANS COMMISSION, 2002), available at: <http://www.pewtrusts.org/pdf/environment_pew_oceans_effects_fishing.pdf> (hereinafter EFFECTS OF FISHING); Gabriella Bianchi, et al., *Impact of fishing on size, composition and diversity of demersal fish communities*, 57 ICES J. MARINE SCIENCE 558, 570 (2000) (analyzes the evidence that fishing, including trawling along the ocean floor, drives down the size and diversity of fish in deep ocean waters).

99. *See* Daniel Pauly, et al., *Fishing down marine food webs*, 279 SCIENCE 860, 860–63 (1998), available at: <http://www.fisheries.ubc.ca/members/dpauly/Science_6_Feb_1998.htm>.

100. *See generally* FAO, *The state of the world's fisheries and aquaculture 2000*, Figure 8 (2000), available at: <http://www.fao.org/sof/sofia/index_en.htm> (hereinafter *World's fisheries*). These "SOFIA" reports are issued every two years, and the 2002 and 2004 reports likewise do not reflect much recent cause for optimism. *See also* Felicia C. Coleman and Susan

L. Williams, *Overexploiting marine ecosystem engineers: potential consequences for biodiversity*, 17 TRENDS IN ECOLOGY & EVOLUTION 40–42 (2002), available at: <http://www.bio.fsu.edu/mote/colemanTREE_01.02.pdf> (describes trophic cascades caused by fishing down the food web, including loss of vitally important species known as "ecosystem engineers" that either morphologically or behaviorally create more complex, and hence more ecologically useful, habitat).

101. *See, e.g.*, Carlton, et al., *supra* note 88, at 517–20 (describes challenges inherent in establishing which marine species have become extinct).

102. The French deep-sea submersible is called *Nautile*, and the Russian vessels are known as MIR-1 and MIR-2. The Russian submersibles can penetrate to approximately 19,800 feet below the surface. *See* Jennifer Uscher, *Deep-sea machines*, available at: <http://www.pbs.org/wgbh/nova/abyss/frontier/deepsea2.html>.

103. *See* NOAA Ocean Explorer, *Alvin*, available at: <http://ocean explorer.noaa.gov/technology/subs/alvin/alvin.html>.

104. *See* The Ocean Conservancy, HEALTH OF THE OCEANS, 2002 REPORT, 17–18 (Ocean Conservancy, 2002) (hereinafter HEALTH OF THE OCEANS).

105. *Id.* at 17.

106. Ransom A. Myers and Boris Worm, *Rapid worldwide depletion of predatory fish communities*, 423 NATURE 280, 282 (May 15, 2003). *See also* supplementary materials related to this article at: <http://ram.biology.dal.ca/depletion/> and <http://ram.biology.dal.ca/depletion/docs/Myers WormFinalPR.pdf>. These materials present additional newly available evidence that only 10 percent of all large fish are now left in the global ocean, compared with their peak numbers of the recent past.

107. *Id.* at 280.

108. *Id.* at 282; HEALTH OF THE OCEANS, *supra* note 104, at 18–19; Ransom A. Myers and Boris Worm, *Extinction, survival or recovery of large predatory fishes*, PHILOSOPHICAL TRANSACTIONS OF THE ROYAL SOCIETY: BIOLOGICAL SCIENCES vol. 360, no. 1453, 13–20 (January 29, 2005).

109. Ove Hoegh-Guldberg, *Climate change, coral bleaching, and the future of the world's coral reefs*, 50 MARINE AND FRESHWATER RESEARCH 839 (1999).

110. *Id. See also Sea of troubles*, *supra* note 86, at 15.

111. *See* Paul K. Dayton, et al., *Environmental effects of marine fishing*, 5 AQUATIC CONSERVATION 205, 206–19 (1995).

112. *World's fisheries*, *supra* note 100.

113. *See* Lester Brown, et al., VITAL SIGNS 2000, 40–41 (W.W. Norton, 2000); WORLD'S FISHERIES, *supra* note 100; J.R. McGoodwin, CRISIS IN THE WORLD'S FISHERIES: PEOPLE, PROBLEMS, AND POLICIES, 51 (Stanford University Press, 1990).

114. *See* Rick Weiss, *Fishing has decimated major species, study says*, WASHINGTON POST (May 14, 2003).

115. *Id.*

116. *Id.* For example, in the Gulf of Thailand, 60 percent of the large finfish vanished in just the first five years of industrialized trawl fishing during the 1960s.

117. Muro-ami is a crude technique in use in the Philippines utilizing stones, chains, and poles to break up coral and induce fish to swim into nets. *See* Michel J. Kaiser, et al., *Modification of marine habitats by trawling activities: Prognosis and solutions*, 3 FISH AND FISHERIES 114, 116 (2002).

118. *See* Carlton, et al., *supra* note 88, at 531; Chuenpagdee, et al., *supra* note 95; National Research Council, EFFECTS OF TRAWLING AND DREDGING ON SEAFLOOR HABITAT, 12–17 (National Academy Press, 2002), available at: <http://books.nap.edu/books/0309083400/html/index .html> (hereinafter EFFECTS OF TRAWLING).

119. *See* Dayton, et al., *supra* note 111, at 206–10; Chuenpagdee, et al., *supra* note 95; John G. Pope, et al., *Gauging the impact of fishing mortality on non-target species*, 57 ICES JOURNAL OF MARINE SCIENCE 689, 693–95 (2000), available at: <http://www.uea.ac.uk/bio/reynoldslab/documents/ Pope_et_al._IJMS_00.pdf>. *See generally* Michel J. Kaiser and Sebastiann J. de Groot, eds., EFFECTS OF FISHING ON NON-TARGET SPECIES AND HABITATS (Blackwell Science, 2000).

120. *See* Carl Safina, *World's imperiled fish (global fish declines)*, 273 SCIENTIFIC AMERICAN 46–53 (1995), available in expanded form at: <http://www.seaweb.org/background/safina6.html>.

121. *See* Colin Woodard, OCEAN'S END: A TRAVEL THROUGH ENDANGERED SEAS, 43–44 (Basic Books, 2000). *See also* HEALTH OF THE OCEANS, *supra* note 104, at 20 (places the by-catch estimate at one-quarter of the annual global fish catch of 84 million tons); EFFECTS OF FISHING, *supra* note 98, at 16–23; U.S. Commission on Ocean Policy, *An ocean blueprint for the 21st century, final report, Washington, D.C.*, Ch. 19 (2004), available at: <http://www.oceancommission.gov/documents/full_color_rpt/000_ ocean_full_report.pdf>.

122. Students will recognize that this is the only known example wherein "high grading" takes on a negative connotation.

123. Because they are made of synthetic organic materials (nylon/plastic), drift nets can be much, much larger than any other nets in history. Their materials are very persistent in the water, lingering as deadly ghost nets that refuse to biodegrade long after people stop checking on them. Also, they can neither be seen nor heard by many marine life forms because they are beyond the detection capabilities of the sonarlike echolocation systems used by some marine mammals. Thus, modern drift nets are far more harmful to marine biodiversity than the smaller, more easily detectable, more biodegradable nets used for millennia. *See* James Carr and Matthew Gianni, *High seas fisheries, large-scale drift nets, and the law of the sea*, 274–76, in FREEDOM FOR THE SEAS IN THE 21ST CENTURY (Jon Van Dyke, et al., eds., Island Press, 1993).

124. *Id. See also* European Cetacean Bycatch Campaign, *Drift nets,* available at: <http://www.eurocbc.org/page357.html>; Roger Payne, *Getting caught in a drift net off Sri Lanka,* VOICE FROM THE SEA (April 19, 2000), available at: <http://www.pbs.org/odyssey/voice/20000419_vos_transcript.html>; Rebecca L. Lewison, et al., *Quantifying the effects of fisheries on threatened species: the impact of pelagic longlines on loggerhead and leatherback sea turtles,* 7 ECOLOGY LETTERS 221–31 (2004), available at: <http://www.k-state.edu/bsanderc/avianecology/lewison2004.pdf>.

125. *See* American Bird Conservancy, *Stopping seabird bycatch. Longline fishing: A global crisis for seabirds—working for solutions to benefit seabirds and fishermen* (2005), available at: <http://www.abcbirds.org/policy/seabird_report_southern.pdf> (concentrates on the Southern hemisphere); American Bird Conservancy, *Sudden death on the high seas. Longline fishing: A global catastrophe for seabirds* (2001), available at: <http://www.abcbirds.org/policy/seabird_report.PDF> (focuses on the Northern hemisphere).

126. *See U.N. General Assembly Resolution 46/215 on large-scale pelagic drift-net fishing and its impact on the living marine resources of the world's oceans and seas,* U.N. Doc. A/RES/46/215 (1991), available at: <http://ods-dds-ny.un.org/doc/RESOLUTION/GEN/NR0/583/03/IMG/NR058303.pdf?OpenElement>. This resolution called for a 50 percent reduction in drift net by mid-1992. *See* Lewison, et al., *supra* note 124, at 225–26 (presenting evidence that some 200,000 loggerheads and 50,000 leatherbacks were trapped globally by longlines in the year 2000 alone, with significant numbers of fatalities, thereby contributing to the catastrophic 80–95 percent loss of these populations in the Pacific during the last twenty years); Larry B. Crowder and Ransom A. Myers, *A comprehensive study of the ecological impacts of the worldwide pelagic longline industry, Report to the Pew Charitable Trusts* (2001), available at: <http://www.seaturtles.org/pdf/Pew_Longline_ 2002.pdf>

(analyzes in detail the still-enormous magnitude of the current problem of longline fishing, with over 5 million baited hooks set each day on 100,000 miles of line worldwide). As the use of drift nets has declined, more and lengthier longlines have more than made up for any loss of killing power, to devastating effect on larger turtles and other unintended victims. *Id.*

127. *See* Carlton, et al., *supra* note 88, at 531; Chuenpagdee, et al., *supra* note 95; *Deep sea corals*, *supra* note 83, at 10–13 (describes the incalculable harm trawling does to vital deep-sea corals, which can take many centuries to form and only minutes to ruin); Jason Hall-Spencer, et al., *Trawling damage to Northeast Atlantic ancient coral reefs*, 269 PROC. ROYAL SOC. BIOLOGICAL SCI. No. 1490, 507–11 (2002).

128. *Deep sea corals*, *supra* note 83, at 12.

129. *See generally* Marine Conservation Biology Institute for the Deep Sea Conservation Coalition, *Debunking claims of sustainability: High-seas bottom trawl red-herrings* (April 2005), available at: <http://www.savethehighseas. org/publicdocs/DSCC_RedHerrings.pdf>.

130. *See* Kaiser, et al., *supra* note 117, at 116–22. *See also* Jeremy S. Collie, et al., *A quantitative analysis of fishing impacts on shelf-sea benthos*, 69 J. ANIMAL ECOL. 785, 793–95 (2000).

131. *See* Michel J. Kaiser, et al., *Fishing-gear restrictions and conservation of benthic habitat complexity*, 14 CONSERVATION BIOLOGY 1512, 1513–14 (2000); Kaiser, *supra* note 117, at 119–22; EFFECTS OF TRAWLING, *supra* note 118, at 18–29.

132. *See* Carlton, et al., *supra* note 88, at 531. The effect of dragging (trawling and dredging) has become far more severe as the increased use of mobile gear such as rollers, rock-hoppers, streetsweepers, precision depth finders, Global Positioning Systems, and more powerful engines has exacted a terrible toll on continental shelves. The phenomenon has been likened to the clear-cutting of tropical forests because of its wholesale and indiscriminate elimination of ecological niches and critical habitats. *See also* Les Watling and Elliott A. Norse, *Disturbance of the seabed by mobile fishing gear: A comparison to forest clearcutting*, 12 CONSERVATION BIOLOGY 1180, 1191–94 (1998), available at: <http://www.stir.ac.uk/Departments/NaturalSciences/DBMS/coursenotes/ 28k7/bottom%20trawl.pdf> (describes the devastating effects of trawling on the seabed, including wholesale crushing, burying, and exposing benthic organisms to predation, and altering sediment and water-column biochemistry).

133. *See* L.A. Robinson and C.L.J. Frid, *Dynamic ecosystem models and the evaluation of ecosystem effects of fishing: Can we make meaningful predictions?* 13 AQUATIC CONSERV. MAR. FRESHW. ECOSYST. 5, 6, 9–10 (2003),

available at: <http://www.environmental-center.com/magazine/wiley/ 1052–7613/pdf2.pdf>.

134. *Sea of troubles, supra* note 86, at 12.

135. The term "tragedy of the commons" was coined by Garrett Hardin and has since been used, often inaccurately, countless times. Garrett Hardin, *The tragedy of the commons*, 162 SCIENCE 1244 (1968). This is how Hardin described the problem:

> Picture a pasture open to all. It is to be expected that each herdsman will try to keep as many cattle as possible on the commons. Such an arrangement may work reasonably satisfactorily for centuries because tribal wars, poaching, and disease keep the numbers of both man and beast well below the carrying capacity of the land. Finally, however, comes the day of reckoning, that is, the day when the long-desired goal of social stability becomes a reality. At this point, the inherent logic of the commons remorselessly generates tragedy.
>
> As a rational being, each herdsman seeks to maximize his gain. Explicitly or implicitly, more or less consciously, he asks, "What is the utility *to me* of adding one more animal to my herd?" The utility has one negative and one positive component.
>
> 1. The positive component is a function of the increment of one animal. Since the herdsman receives all the proceeds from the sale of the additional animal, the positive utility is nearly +1.
> 2. The negative component is a function of the additional overgrazing created by one more animal. Since, however, the effects of overgrazing are shared by all the herdsmen, the negative utility for any particular decision-making herdsman is only a fraction of −1.
>
> Adding together the component partial utilities, the rational herdsman concludes that the only sensible course for him to pursue is to add another animal to his herd. And another; and another. . . .
>
> But this is the conclusion reached by each and every rational herdsman sharing a commons. Therein is the tragedy. Each man is locked into a system that compels him to increase his herd without limit—in a world that is limited. Ruin is the destination toward which all men rush, each pursuing his own best interest in a society that believes in the freedom of the commons. Freedom in a commons brings ruin to all.

136. Here is how I conceptualize the tragedy of the commons on a level that I can grasp. Such common resources as the oceans can be analogized

to a glass of soda being shared by myself and my young daughters Christie and JulieKate, each of us with a straw. The "rule of capture" is in effect; meaningful ownership of the soda is not established until it is in one's possession (i.e., consumed). The predictable result is that Christie, JulieKate, and I all drink as fast as we are able, until the glass is empty. If I decide to exercise some mature fatherly restraint and stop drinking, or drink more slowly, so as to conserve some soda for later, my self-control would be rewarded only by the spectacle of my daughters swiftly draining the glass dry. *See* Richard Stroup and John Baden quoted in Robert Smith, *Preserving the earth: The property rights approach,* 4 CATO INSTITUTE POLICY REPORT 7, 9 (1982).

137. *See* Carlton, et al., *supra* note 88, at 529–30 (states that the number of species of marine mammals, turtles, and fish that are now severely reduced due to direct hunting appears to be unprecedented in all of human history); HEALTH OF THE OCEANS, *supra* note 104, at 18–19; Watling and Norse, *supra* note 132, at 1181–83, 1187–90 (details the appalling damage to marine biodiversity and key habitats inflicted by various types of trawls).

138. *See Sea of troubles, supra* note 86, at 10–15.

139. *See* LIVING OCEAN, *supra* note 26, at 22.

140. *Id.*

141. *See Sea of troubles, supra* note 86, at 3, 8–9.

142. *See* LIVING OCEAN, *supra* note 26, at 24.

143. *See Sea of troubles, supra* note 86, at 8.

144. *Id.* at 14.

145. *See* World Wildlife Fund (WWF), *Showcase examples for the OSPAR system of marine protected areas (MPAs),* 17, available at: <http://www.ngo .grida.no/wwfneap/Projects/MPAmap.htm#galicia>.

146. David Pinder, *Offshore oil and gas: Global resource knowledge and technological change,* OCEAN AND COASTAL MANAGEMENT vol. 583, no. 9–10, 44 (2001).

147. *Id.*

148. *Id.*

149. *See id.* at 584. The estimated percentages of oil reserves located under the deep ocean floor for three regions are: Latin America (49.2 percent), West Africa (49.9 percent), and North America (59.9 percent).

150. *See* Kevin Krajick, *The crystal fuel,* 106 NAT. HIST. 26, 26–27 (1997).

151. *See* U.S. Dep't of Interior, Minerals Management Service, *An assessment of the undiscovered hydrocarbon potential of the nation's outer continental shelf: A resource evaluation program report,* MMS 96–0034 (1996).

152. *See generally*, Craig H. Allen, *Protecting the oceanic Gardens of Eden: International law issues in deep-sea vent resource conservation and management*, 13 GEO. INT'L ENVT'L L. REV. 563, 576–83 (2001).

153. Office of Naval Research, *Habitats: Hydrothermal vent—hydrothermal vent life*, available at: <www.onr.navy.mil/focus/ocean/habitats/vents2.htm>. One site has been found containing polymetallic sulfides estimated to be worth two billion dollars.

154. *See generally* Allen, *supra* note 152, at 578–81. The U.S. Department of the Interior (DOI) attempted to lease up to 70,000 square miles of the Gorda Ridge area seabed off the Northern California-Oregon coast for mining. The DOI proceeded through an extended National Environmental Policy Act (NEPA) review process, and in the face of stiff resistance to their actions, they concluded that the area would not be a potential target for commercial mining development for several more decades; they then abandoned the initiative.

155. *See id.* at 580.

156. *See* Jan Magne Markussen, *Deep seabed mining and the environment: Consequences, perceptions, and regulations*, 31–39, in GREEN GLOBE YEARBOOK OF INTERNATIONAL CO-OPERATION ON ENVIRONMENT AND DEVELOPMENT 1994 (1994), available at: <http://www.greenyearbook.org/articles/94_02_markussen.pdf> (hereinafter GREEN GLOBE YEARBOOK). Markussen explains that there are three main types of deep-seabed minerals: polymetallic nodules, crusts, and sulfides (31). *See also* Inter-Ridge Report, *Management and Conservation of Hydrothermal Vent Ecosystems* (2000), available at: <http://195.37.14.189/public_html/SCIENCE/Science_reports/ReportPDFs/ventrepMay01.pdf>.

157. GREEN GLOBE YEARBOOK, *supra* note 156, at 33–34, 35; *Deep sea corals*, *supra* note 83, at 13.

158. GREEN GLOBE YEARBOOK, *supra* note 156, at 34.

159. *Id.*

160. *See generally* Allen, *supra* note 152, at 583–85.

161. EFFECTS OF FISHING, *supra* note 98.

162. *See* Rupert F.G. Ormond, John D. Gage, and Martin V. Angel, eds., MARINE BIODIVERSITY: PATTERNS AND PROCESSES, xiii (Cambridge University Press, 1997).

163. *See Offshore directory*, *supra* note 50, at 20. *See also* International Council for the Exploration of the Seas, *Is time running out for deep sea fish?* available at: <http://www.ices.dk/marineworld/deepseafish.asp> (notes that orange roughy are one of the slowest-growing of all deepwater fish and can live to 125 years of age).

164. *See id.* Stocks of orange roughy have been fished down to 15–30 percent of their initial biomass within five to ten years of their discovery. Also, the pelagic armourhead was fished to commercial extinction within ten years of its discovery over the seamounts in the northern Hawaiian Ridge.

165. *See* World Wildlife Fund, *supra* note 145; *WWF urges the European Union to take action against seamount exploitation* (November 12, 2002), available at: <http://www.ngo.grida.no/wwfneap/Publication/pr121102 .htm>.

166. *See id.*

167. The International Council for the Exploration of the Sea (ICES) was created through the Convention for the International Council for the Exploration of the Sea (1964). It is an intergovernmental organization consisting of nineteen member nations. The primary duty of the council is to promote and encourage research and investigations for the study of the sea (Art. 1[a]) and to publish or otherwise disseminate the results of research and investigations carried out under its auspices or to encourage the publication thereof (Art. 1[c]). The advice of the council has no binding authority. *See* ICES website, available at: <http://www.ices.dk/>.

168. *See* World Wildlife Fund, *supra* note 145.

169. *See id.*

170. *Initial OSPAR list of threatened and/or declining species and habitats,* available at: <www.ospar.org>. Now, ICES has acknowledged threats to seamounts from overfishing, extensive trawling activities, and other impacts. *See* Telmo Morato, *Seamounts—hotspots of marine life,* available at: <www .ices.dk/marineworld.seamounts.asp>.

171. *See* Mark Schrope, *60,000 bucks under the sea,* OUTSIDE ONLINE MAGAZINE (June 2000), available at: <http://web.outsideonline.com/ magazine/200006/200006disp2.html>. *See also* Lauren Mullineaux, *Biology working group update,* INTERRIDGE NEWS 10 (fall 1999) (reports that one such company, Aegraham DeepSea Voyages, has already carried a team of ecotourists to the Rainbow hydrothermal vent site on the Mid-Atlantic Ridge).

172. Roberts, et al., *supra* note 82, at 1280.

173. *Id.* at 1281.

174. Brad Phillips, *The ocean's top 10 coral reef hotspots identified for the first time: Study sounds alarm for extinctions of marine species,* Conservation International website (February 14, 2002), available at: <http://www.conservation .org/xp/news/press_releases/2002/021402.xml>.

175. *Id.*

176. *Id.*

177. WWF, *About the "Global 200" ecoregions*, available at: <http://www
.panda.org/about_wwf/where_we_work/ecoregions/about/index.cfm>.

178. *Id.*

179. *Id.* The WWF website has detailed information about each of these
marine eco-regions, including a risk assessment and an explanation as to
why they are so ecologically significant. WWF, *Selection of marine ecoregions*,
available at: <http://www.panda.org/about_wwf/where_we_work/eco
regions/about/habitat_types/selecting_marine_ecoregions/index.cfm>.

180. *See generally* Great Barrier Reef Marine Park Authority, the World
Bank, and the World Conservation Union (Graeme Kelleher, Chris Bleak-
ley, and Sue Wells, eds.), A GLOBAL REPRESENTATIVE SYSTEM OF
MARINE PROTECTED AREAS, VOL. I (The World Bank, Washington,
D.C., 1995).

181. Oceans Act of 2000, PUB. L. 106–256, available at: <http://www
.oceancommission.gov/documents/oceanact.pdf>.

182. U.S. Commission on Ocean Policy, *An ocean blueprint for the 21st
century, final report, Washington, D.C.* (2004), available at: <http://www.ocean
commission.gov/documents/full_color_rpt/000_ocean_full_report.pdf>.

CHAPTER TWO

1. Some of the other global international agreements with significant
relevance to marine biodiversity are: Agreement on Straddling Fish Stocks
and Highly Migratory Fish Stocks (December 1995); International Convention for the Regulation of Whaling (1946); Convention on the Transboundary Movement of Hazardous Wastes (Basel Convention) (1989);
United Nations Framework Convention on Climate Change (1992); and the
Convention on the Conservation of Migratory Species of Wild Animals
(Bonn Convention) (June 1979). There are also nonbinding U.N. agreements such as the Montreal Guidelines for the Protection of the Marine
Environment Against Pollution from Land-Based Sources (May 1985). *See*
Rachel Wynberg, *International and national policies concerning marine and
coastal biodiversity*, MARINE BIODIVERSITY STATUS REPORT (March
2000), available at: <http://www.nrf.ac.za/publications/marinerep/policies
.htm>.

2. Some of the major regional multinational legal agreements relevant
to marine biodiversity include: Helsinki Convention on the Protection of
the Marine Environment of the Baltic Sea Area (1992); OSPAR (Oslo-

Paris) Convention; Convention for the Protection of the Marine Environment of the North-East Atlantic (1998); Convention for Cooperation in the Protection and Development of the Marine and Coastal Environment of the West and Central African Region (Abidjan Convention) (1981); Convention for the Protection of the Marine Environment and Coastal Area of the South-East Pacific (Lima Convention) (1981); Convention for the Protection of Natural Resources and Environment of the South Pacific Region (Noumea Convention) (1986); Convention for the Prohibition of Fishing with Long Driftnets in the South Pacific (1989); Convention for Cooperation in the Protection and Sustainable Development of the Marine and Coastal Environment of the Northeast Pacific (2002); Convention on the Conservation of Antarctic Marine Living Resources (1980); and Convention for the Protection, Management and Development of the Marine and Coastal Environment of the Eastern African Region (Nairobi Convention) (1985). There are protocols attendant to many of these regional conventions as well.

3. The Third U.N. Convention on the Law of the Sea, entered into force in 1994, U.N. Doc. A/CONF.62/122, 21 I.L.M. 1261 (1982). This agreement is commonly known by many names, including UNCLOS, UNCLOS 1982, the Law of the Sea Treaty, LOS Convention, and LOSC. *See* Oceans and Law of the Sea, Division for Ocean Affairs and the Law of the Sea website, available at: <http://www.un.org/Depts/los/index.htm>.

4. Part XII of UNCLOS consists of eleven sections composed of Art. 192–237.

5. *See* David D. Newsom, *The Senate's distaste for treaties—A perennial problem for US diplomacy*, CHRISTIAN SCIENCE MONITOR 19 (January 14, 1998); Paul S. Sochaczewski and Jay Hyvarinen, *Down deep; Macro- and micro-flora and fauna of the deep sea thermal vents*, EARTH ACTION NETWORK vol. 7, no. 4, 15 (July 1996). It is Part XII of UNCLOS (Art. 192–237) that generally covers protection of the marine environment, although the term "marine environment" itself is never defined in the convention.

6. *See* G.P. Glasby, *Lessons learned from deep-sea mining; Sea mineral deposits*, 289 SCIENCE 551 (July 28, 2000).

7. *Id.*

8. *See* Oceans and Law of the Sea, Division for Ocean Affairs and the Law of the Sea, *United Nations Convention on the Law of the Sea, agreement relating to the implementation of Part XI of the Convention*, available at: <http://www.un.org/Depts/los/convention_agreements/convention_agreements.htm>.

9. *See* the list of parties and ratification dates at Oceans and Law of the Sea, Division for Ocean Affairs and the Law of the Sea, *Chronological lists of ratifications of, accessions and succession to the Convention and the related Agreements as at 20 September 2005*, available at: <http://www.un.org/Depts/los/reference_files/chronological_lists_of_ratifications.htm#The%20United%20Nations%20Convention%20on%20the%20Law%20of%20the%20Sea>.

10. Certain aspects of the jurisdictional regime described in this chapter are also commonly considered to be customary international law. For instance, "freedom of the high seas" is a concept that dates back to Hugo Grotius in the early seventeenth century, when Grotius opined that "the sea is common to all because it is so limitless that it cannot become a possession of one, and because it is adapted for the use of all, whether we consider it from the point of view of navigation or of fisheries." *See* R.P. Anand, *Changing concepts of freedom of the seas: A historical perspective*, 74–75, in FREEDOM FOR THE SEAS IN THE 21ST CENTURY (Jon Van Dyke, et al., eds., Island Press, 1993). The concepts of the continental shelf and exclusive economic zone (EEZ) are also recognized by some nations under customary international law. *See* Biliana Cicin-Sain and Robert W. Knecht, THE FUTURE OF U.S. OCEAN POLICY, 34 (Island Press, 2000). The United States has not ratified UNCLOS, but has nevertheless declared a 200 nautical mile EEZ and continental-shelf rights which did not exist prior to UNCLOS. On March 10, 1983, President Reagan issued a Statement on United States Oceans Policy in which he explained that the United States refused to sign because of the deep-seabed mining provisions, but he also said that UNCLOS "contains provisions with respect to traditional uses of the oceans which generally confirm existing maritime law and practice and fairly balance the interests of all parties." The president's statement also averred that "the United States is prepared to accept and act in accordance with the balance of interests relating to traditional uses of the ocean." *See* Ronald Reagan, *Statement on United States ocean policy, March 10, 1983*.

11. In the United States, most individual states manifest sovereignty over submerged lands and overlying waters to a distance of three miles. *See* Carol E. Remy, *U.S. territorial sea extension: Jurisdiction and international environmental protection*, 16 FORD. INT'L L. J. 1208, 1210 (1993).

12. *Third United Nations Convention on the Law of the Sea* (October 7, 1982), reprinted in 21 I.L.M. 1261, Art. 3 (1982) (hereinafter UNCLOS). Subject to the right of innocent passage (Art. 17).

13. UNCLOS, Art. 245.

14. UNCLOS, Art. 17.

15. UNCLOS, Art. 21(1)(d).

16. UNCLOS, Art. 21(1)(f).

17. UNCLOS, Art. 57. The EEZ does not automatically exist; a nation must expressly declare it (Art. 56[1][a]).

18. UNCLOS, Art. 56(1)(a).

19. UNCLOS, Art. 62.

20. UNCLOS, Art. 61. Limitations include straddling stocks, highly migratory species, catadromous and anadromous species (Art. 63–67).

21. UNCLOS, Art. 61(1–3).

22. *See* Barbara Kwiatkowska, THE 200 MILE EXCLUSIVE ECONOMIC ZONE IN THE NEW LAW OF THE SEA, xxiii (Springer, 1989).

23. UNCLOS, Art. 61(4).

24. UNCLOS, Art. 56(1)(b)(ii) and (iii), and Art. 246. This control over marine scientific research has important implications. Many developing nations view marine scientific research as heavily benefiting developed nations. *See* Craig H. Allen, *Protecting the oceanic Gardens of Eden: International law issues in deep-sea vent resource conservation and management*, 13 GEO. INT'L ENVT'L L. REV. 563, 587–88 (2001). Consequently, UNCLOS requires coastal state approval for conducting research within the EEZ. This requirement can hamper efforts to locate these important areas, and a coastal nation intent on exploiting the resources in and around these areas may block research efforts in order to prevent the resources from being discovered.

25. UNCLOS, Art. 58.

26. UNCLOS, Art. 62.

27. UNCLOS, Art. 246(3).

28. UNCLOS, Art. 77–81.

29. UNCLOS, Art. 76(1), (3–6). Art. 76(1), (3–6) reads:

    1. The continental shelf of a coastal State comprises the seabed and subsoil of the submarine areas that extend beyond its territorial sea throughout the natural prolongation of its land territory to the outer edge of the continental margin, or to a distance of 200 nautical miles from the baselines from which the breadth of the territorial sea is measured where the outer edge of the continental margin does not extend up to that distance.

    3. The continental margin comprises the submerged prolongation of the land mass of the coastal State, and consists of the seabed and subsoil of the shelf, the slope and the rise. It does not include the deep ocean floor with its oceanic ridges or the subsoil thereof.

4. (a) For the purposes of this Convention, the coastal State shall establish the outer edge of the continental margin wherever the margin extends beyond 200 nautical miles from the baselines from which the breadth of the territorial sea is measured, by either:

   (i) a line delineated in accordance with paragraph 7 by reference to the outermost fixed points at each of which the thickness of sedimentary rocks is at least 1 per cent of the shortest distance from such point to the foot of the continental slope; or

   (ii) a line delineated in accordance with paragraph 7 by reference to fixed points not more than 60 nautical miles from the foot of the continental slope.

   (b) In the absence of evidence to the contrary, the foot of the continental slope shall be determined as the point of maximum change in the gradient at its base.

5. The fixed points comprising the line of the outer limits of the continental shelf on the seabed, drawn in accordance with paragraph 4 (a)(i) and (ii), either shall not exceed 350 nautical miles from the baselines from which the breadth of the territorial sea is measured or shall not exceed 100 nautical miles from the 2,500 meter isobath, which is a line connecting the depth of 2,500 meters.

6. Notwithstanding the provisions of paragraph 5, on submarine ridges, the outer limit of the continental shelf shall not exceed 350 nautical miles from the baselines from which the breadth of the territorial sea is measured. This paragraph does not apply to submarine elevations that are natural components of the continental margin, such as its plateaux, rises, caps, banks and spurs.

30. UNCLOS, Art. 78(1).

31. UNCLOS, Art. 74(4). *See also* UNCLOS, Art. 81 which reads, "The coastal State shall have the exclusive right to authorize and regulate drilling on the continental shelf for all purposes." Under the original UNCLOS (1982) the coastal State was required to make payments, or in-kind contributions, of 7 percent of the value of nonliving resources produced from the area of the continental shelf beyond 200 nautical miles. These payments would then be distributed to state parties to the convention on an "equitable sharing" basis (UNCLOS, Art. 82). Several nations, including the United States, objected to these provisions and refused to sign or ratify the convention. These provisions were eventually changed to reduce the payments and to eliminate mandatory technology transfers. *See* Oceans and Law of the

Sea, Division for Ocean Affairs and the Law of the Sea, *United Nations Convention on the Law of the Sea, Agreement Relating to the Implementation of Part XI of the Convention*, available at: <www.un.org/Depts/los/convention_agree ments.convention_agreements.htm>.

32. *Id.*

33. UNCLOS, Art. 77(3).

34. UNCLOS, Art. 78.

35. UNCLOS, Art. 79.

36. UNCLOS, Art. 77 and 87.

37. UNCLOS, Art. 86.

38. UNCLOS, Art. 87. Activities include fishing, navigation, overflight, laying of cables and pipelines, and scientific research. Freedom of fishing is subject to limitations. Fishing for straddling stocks, highly migratory species, and catadromous and anadromous species is prohibited (Art. 63–67). All states enjoy the right to conduct marine scientific research (Art. 87[1][f]). Research on the deep seabed of the high seas—defined by UNCLOS as "the Area" (Art. 1[1][1])—is subject to the requirement that such activities be carried out for peaceful purposes and for the common benefit of all mankind (Art. 143[1]).

39. UNCLOS, Art. 89.

40. UNCLOS, Art. 136.

41. Sochaczewski and Hyvarinen, *supra* note 5.

42. UNCLOS, Art. 1(1)(1).

43. UNCLOS, Art. 133. It is important to note that the marine biodiversity, including deep-seabed genetic resources, would remain outside of the regime set up to govern the international seabed area.

44. *Id.* The "Authority" is the International Seabed Authority (Art. 1[2]). For "common benefit of mankind," *see* UNCLOS, Art. 140. For "peaceful purposes," *see* UNCLOS, Art. 141.

45. UNCLOS, Art. 117–19.

46. UNCLOS, Art. 87(1)(e).

47. I say "presumably" because even while declaring the freedom to fish on the high seas, Art. 87(1)(e) states that this freedom is "subject to the conditions laid down in section 2." Section 2 obligates parties to enact and cooperate in enacting regulations to conserve and manage the "living resources of the high seas." *See* UNCLOS, Art. 117–19.

48. *See* Lyle Glowka, *The deepest of ironies: Genetic resources, marine scientific research, and the area*, 12 OCEAN Y.B. 154, 168–69 (1996). From this viewpoint, the unique microbial genetic diversity located at deep-seabed hydrothermal vents would fall outside of these conservation provisions.

49. *See, e.g.*, Lawrence Juda and R.H. Burroughs, *The prospects for comprehensive ocean management*, MARINE POLICY 31 (January 1990). *See also* Pew Oceans Commission, *America's living oceans: Charting a course for sea change*, 26 (May 2003), available at: <http://www.oceanconservancy.org/site/DocServer/oceans_report.pdf?docID=242> (hereinafter *Sea change*) (describes the formulation of U.S. ocean policy as being predicated on a sector-by-sector, crisis-by-crisis basis). *See also Sea change*, at 27 (relates that over 140 U.S. laws pertain in some respect to the ocean and coast).

50. *See, e.g.*, Donald K. Anton, *Law for the sea's biological diversity*, 36 COLUM. J. TRANSNAT'L L. 341, 362 (1997) (notes that fifty-plus treaties and instruments for the protection of the marine environment have been negotiated since 1972).

51. *See* Terence P. Stewart and David S. Johanson, *A nexus of trade and the environment: The relationship between the cartagena protocol on biosafety and the SPS agreement of the World Trade Organization*, 14 COLO. J. INT'L ENVTL. L. & POL'Y 1, 40 (Winter 2003).

52. This is from the delineation in the Rio Declaration on Environment and Development of 1992, Principle 15. *See Rio Declaration on Environment and Development, June 14, 1992*, UN Doc. A/CONF.151/5/Rev.1 (1992), available at: <http://www.unep.org/Documents.multilingual/Default.asp?DocumentID=78&ArticleID=1163>. *See also* Catherine Tinker, *Is a United Nations convention the most appropriate means to pursue the goal of biological diversity?: Responsibility for biological diversity conservation under international law*, 28 VAND. J. TRANSNAT'L L. 777, 779 (October 1995) (states that there is no agreement on the content of the precautionary principle).

53. *See* Arie Trouwborst, EVOLUTION AND STATUS OF THE PRECAUTIONARY PRINCIPLE IN INTERNATIONAL LAW (The Hague; London; Boston, Kluwer Law International, 2002). Trouwborst concludes that "the precautionary principle is not only a general, perhaps even universal custom in that it binds, in principle, all governments of the world, but also in that it aims for comprehensive environmental protection" (284). *See also WTO Appellate Body Report on EC Measures Concerning Meat and Meat Products (Hormones)*, WT/DS26/AB/R, WT/DS48/AB/R, AB-1997-4, para. 121 (January 16, 1998) (states that the precautionary principle "at least outside the field of international environmental law, still awaits authoritative formulation" as customary international law).

54. *See, e.g., United Nations Framework Convention on Climate Change (FCCC), May 9, 1992*, 31 ILM 849 (1992); *Convention on Biological Diversity, June 5, 1992*, 31 ILM 818 (1992); *Kyoto Protocol to the FCCC, Dec. 10, 1997*, 37

ILM 22 (1998); *Cartagena Protocol on Biosafety to the Convention on Biological Diversity, Jan. 29, 2000*, 39 ILM 1027 (2000).

55. *See* Stewart and Johanson, *supra* note 51, at 40–44 (argues that the precautionary principle may not have attained customary international law status due primarily to the lack of a sufficient time lapse; points out that none of the major international tribunals have yet ruled on this point). *See also* Tinker, *supra* note 52, at 795 (argues that the precautionary principle may not have attained customary international law status due to the fact that the international instruments that have included this principle are neither binding nor intended to be binding upon the parties); John D. Graham, *The role of precaution in risk assessment and management: An American's view*, address before the European Commission, the U.S. Mission to the E.U., the German Marshall Fund with the European Policy Centre and the Center for Environmental Solutions (January 11–12, 2002).

56. *Principle 21 of the Stockholm Conference on the Human Environment, Stockholm Declaration of the United Nations Conference on the Human Environment (Principle 21), June 16, 1972*, U.N. Doc. A/Conf.48/14/Rev. 1, 11 I.L.M. 1416, available at: <http://www.unep.org/Documents.multilingual/Default.asp?DocumentID=97&ArticleID=1503>. Principle 21 of the Stockholm Declaration is widely considered as having become a rule of customary international law. *See, e.g.*, Alexandre Kiss and Dinah Shelton, INTERNATIONAL ENVIRONMENTAL LAW, 106–7 (Transnational Publishers, 1991); Tinker, *supra* note 52, at 806 (asserts that this principle made the leap to hard law when it was included in the U.N. Convention on Biological Diversity).

57. *Principle 21.*

58. States have "the sovereign right to exploit their own resources pursuant to their own environmental and developmental policies." *See* Rio Declaration, Principle 2.

59. There is some debate over whether the EEZ and continental-shelf regimes are creatures of customary law or whether they spring forth from UNCLOS. It seems that if these regimes are customary law, then the precautionary principle and Principle 21 would apply as customary international law to these areas as well. The problem then remains enforcement because violation of the precautionary principle at present does not constitute a breach of international law.

60. *See* UNCLOS, Art. 193.

61. *See* Tinker, *supra* note 52, at 794.

62. *Id.*

63. *Id.*

64. *Id.* at 791.

65. *Id.* at 800.

66. *Id.*

67. *See* The National Environmental Policy Act, 42 U.S.C. 4331–4370f, available at: <http://ceq.eh.doe.gov/nepa/regs/nepa/nepaeqia.htm>.

68. *Id.*

69. *See generally* Matthew J. Lindstrom, *Procedures without purpose: The withering away of the National Environmental Policy Act's Substantive Law*, 20 J. LAND RESOURCES & ENVT'L L. 245 (2000).

70. *Citizens to Preserve Overton Park, Inc. v. Volpe*, 401 U.S. 402 (1971).

71. UNCLOS, Art. 117.

72. There is no express legal authority to designate MPAs in the high seas. *See* Allen, *supra* note 24, at 598.

73. *General Agreement on Tariffs and Trade 1994*, 33 I.L.M. 1125 (April 15, 1994), available at: <http://www.wto.org/english/docs_e/legal_e/06-gatt .pdf>. *See* Edith Brown Weiss, *Understanding compliance with international environmental agreements: The baker's dozen myths*, 32 U. RICH. L. REV. 1555, 1585–86 (1999). *See generally* Chris Wold, *Multilateral environmental agreements and the GATT: Conflict and resolution?* 26 ENVT'L L. 841 (1996); Christine Crawford, *Conflicts between the Convention on International Trade in Endangered Species and the GATT in light of actions to halt the rhinoceros and tiger trade*, 7 GEO. INT'L ENVT'L L. REV. 555 (1995).

74. Great Barrier Reef Marine Park Authority, the World Bank, and the World Conservation Union (Graeme Kelleher, Chris Bleakley, and Sue Wells, eds.), A GLOBAL REPRESENTATIVE SYSTEM OF MARINE PROTECTED AREAS, VOL. I (Washington, D.C., World Bank, 1995) (hereinafter GLOBAL SYSTEM).

75. *Id.*

76. *Id.* at 14, 18–19. *See generally* Benjamin Halpern, *The impact of marine reserves: Do reserves work and does reserve size matter?* 13 ECOLOGICAL AP-PLICATIONS 117 (2003), available at: <http://www.nceas.ucsb.edu/~halpern/pdf/Halpern_EA_2003.pdf> (demonstrates that marine reserves do work when appropriately sited, and that larger reserves are verifiably more effective than smaller ones); John Charles Kunich, ARK OF THE BROKEN COVENANT: PROTECTING THE WORLD'S BIODIVERSITY HOTSPOTS, 145–46 (Praeger, 2003).

77. *See* World Wildlife Fund, *Marine reserves: Protecting the future of our oceans*, 4, available at: <www.wwfus.org/oceans/marine_reserves.pdf.> (hereinafter *Marine reserves*). *See also* Callum Roberts, *Marine protected areas as*

strategic tools, COMMUNITY RESEARCH AND DEVELOPMENT IN-FORMATION SERVICE—CORDIS, ACP-EU FISHERIES REPORT NUMBER 5 (1999), available at: <http://www.cordis.lu/inco2/src/acprep26 .htm>. Roberts' report describes several studies that demonstrate the increased productivity of several species of rockfish located within marine preserves. Increased productivity ranged from 2 to 100 times greater than for rockfish located in fished areas.

78. *See Marine reserves, supra* note 77, at 5.

79. *Id.*

80. *Id. See also Sea change, supra* note 49, at 31 (details the benefits of marine reserves, including their ability to restore ecosystems and enhance populations by increasing abundance, diversity, and productivity of marine organisms. Marine reserves also protect the structure and functioning of marine ecosystems and habitats).

81. *See* Roberts, *supra* note 77.

82. *Id.* In the United States 4.6 percent of the land area is protected as reserves, while a "fraction" of 1 percent of the ocean environment is protected as reserves. Some scientists have recommended that 20 percent of the world's oceans should be set aside as no-take zones by the year 2020.

83. *Marine reserves, supra* note 77, at 3.

84. *Id.* at 6.

85. The Florida Keys National Marine Sanctuary is considered an excellent example of a well-managed marine sanctuary. The sanctuary employs a system of ocean zoning in which core conservation areas are protected from all disruptive human activities, while in other areas limited activities, consistent with conservation goals, are permitted.

86. *See* Stephen Lutter and Bernd Christiansen, *WWF urges the European Union to take action against seamount exploitation: Hamburg, Germany, 12 November 2002,* available at: <http://www.ngo.grida.no/wwfneap/Publication/ pr121102.htm>. WWF commended Australia, New Zealand, and Canada for taking the first steps toward protecting seamounts in their offshore waters despite little available data evidencing human impact.

87. *See* National Ocean and Atmospheric Association, *Ocean Explorer,* available at: <http://oceanexplorer.noaa.gov/explorations/02davidson/back ground/missionplan/plan.html>.

88. *See* Tasmanian Seamounts Marine Reserve, available at: <http:// www.ea.gov.au/coasts/mpa/seamounts/plan/pubs/plan.pdf>. The reserve consists of over seventy seamounts. The objective is to preserve the areas in their natural condition. This reserve was established under the Environmental

Protection and Biodiversity Conservation Act of 1999. The stated purpose of the reserve is the full protection of benthic biodiversity, and all fishing, mining, petroleum, and mineral exploration are prohibited.

89. *See, e.g.,* Richard Pollnac, Brian Crawford, and Maharlina L.G. Gorospe, *Discovering factors that influence the success of community-based marine protected areas in the Visayas, Philippines,* 44 OCEAN AND COASTAL MANAGEMENT 683–84 (2001) (states that only 20–25 percent of the over 400 MPAs in the Philippines are successful).

90. *See* Allen, *supra* note 24, at 598.

91. *See* Claudia E. Mills and James T. Carlton, *Rationale for a system of international reserves for the open ocean,* 12 CONSERVATION BIOLOGY 244 (1998), available at: <http://faculty.washington.edu/cemills/ConsBiol1998 .pdf> (analyzes the need for a system of open-ocean reserves with regions where there is no commercial shipping activity, so as to minimize the discharge and spill of wastes and oil; such a system would allow little or no fishing, no deep-sea mining, no dumping, no deployment or testing of weapons, and no "floating cities").

92. *See* UNCLOS Preamble and Art. 4.

93. UNCLOS, Art. 192.

94. UNCLOS, Art. 197.

95. UNCLOS, Art. 194(5).

96. UNCLOS, Art. 62.

97. UNCLOS, Art. 61(4).

98. UNCLOS, Art. 194(1).

99. *Id.*

100. *See* UNCLOS, Art. 210(5), which states:

> Dumping within the territorial sea and the exclusive economic zone or onto the continental shelf shall not be carried out without the express prior approval of the coastal State, which has the right to permit, regulate and control such dumping after due consideration of the matter with other States which by reason of their geographical situation may be adversely affected thereby.

See also Art. 210 (implicitly incorporates the London Dumping Convention with the requirement that national legislation be no less effective than global standards).

101. UNCLOS, Art. 235(1).

102. UNCLOS, Art. 297(1)(c).

103. UNCLOS, Art. 310.

104. An exception to this would be the Straddling Stocks Agreement.

105. The United Nations Agreement for the Implementation of the Provisions of the United Nations Convention on the Law of the Sea relating to the Conservation and Management of Straddling Fish Stocks and Highly Migratory Fish Stocks entered into force on December 11, 2001, after being signed and ratified by thirty-four countries, including the United States. It now has fifty-three ratifications. Available at: <http://www.un.org/Depts/los/refer ence_files/chronological_lists_of_ratifications.htm> (hereinafter Agreement on Straddling Stocks).

106. Agreement on Straddling Stocks, Art. 6(1).

107. Agreement on Straddling Stocks, Art. 6(3)(a).

108. Agreement on Straddling Stocks, Art. 2 and Art. 5 (a) and (b).

109. Agreement on Straddling Stocks, Art. 5(e).

110. Agreement on Straddling Stocks, Art. 5(g).

111. *See* Anton, *supra* note 50, at 144.

112. Convention on Biological Diversity of the United Nations Conference on the Environment and Development, June 5, 1992, U.N. Doc. DPI/1307, reprinted in 31 I.L.M. 818 (1992) (entered into force December 29, 1993) (hereinafter CBD).

113. See the list of parties and ratification dates at Convention on Biological Diversity, *Parties to the Convention on Biological Diversity/Cartagena Protocol on Biosafety*, available at: <http://www.biodiv.org/world/parties.asp>.

114. CBD, Art. 1.

115. *See* Francoise Burhenne-Guilmin and Susan Casey-Lefkowitz, *The Convention on Biological Diversity: A hard won global achievement,* 3 Y.B. INT'L ENVT'L L. 43, 45–46 (1992).

116. CBD, Art. 6.

117. CBD, Art. 7.

118. CBD, Annex I.

119. CBD, Art. 8. Article 8 reads, in its entirety, as follows:

Each Contracting Party shall, as far as possible and as appropriate:

(a) Establish a system of protected areas or areas where special measures need to be taken to conserve biological diversity;

(b) Develop, where necessary, guidelines for the selection, establishment and management of protected areas or areas where special measures need to be taken to conserve biological diversity;

(c) Regulate or manage biological resources important for the conservation of biological diversity whether within or outside protected areas, with a view to ensuring their conservation and sustainable use;

(d) Promote the protection of ecosystems, natural habitats and the maintenance of viable populations of species in natural surroundings;

(e) Promote environmentally sound and sustainable development in areas adjacent to protected areas with a view to furthering protection of these areas;

(f) Rehabilitate and restore degraded ecosystems and promote the recovery of threatened species, inter alia, through the development and implementation of plans or other management strategies;

(g) Establish or maintain means to regulate, manage or control the risks associated with the use and release of living modified organisms resulting from biotechnology which are likely to have adverse environmental impacts that could affect the conservation and sustainable use of biological diversity, taking also into account the risks to human health;

(h) Prevent the introduction of, control or eradicate those alien species which threaten ecosystems, habitats or species;

(i) Endeavor to provide the conditions needed for compatibility between present uses and the conservation of biological diversity and the sustainable use of its components;

(j) Subject to its national legislation, respect, preserve and maintain knowledge, innovations and practices of indigenous and local communities embodying traditional lifestyles relevant for the conservation and sustainable use of biological diversity and promote their wider application with the approval and involvement of the holders of such knowledge, innovations and practices and encourage the equitable sharing of the benefits arising from the utilization of such knowledge, innovations and practices;

(k) Develop or maintain necessary legislation and/or other regulatory provisions for the protection of threatened species and populations;

(l) Where a significant adverse effect on biological diversity has been determined pursuant to Article 7, regulate or manage the relevant processes and categories of activities; and

(m) Cooperate in providing financial and other support for in-situ conservation outlined in subparagraphs (a) to (l) above, particularly to developing countries.

120. CBD, Art. 10.

121. *See* Tinker, *supra* note 52, at 813. Principle 21 establishes the duty of nations not to harm the territory of other states or any territory beyond national jurisdiction (779).

122. CBD, Art. 3.

123. *See* Tinker, *supra* note 52, at 813.

124. This is called for in CBD, Art. 21.

125. CBD, Art. 20.

126. *See* United Nations Development Programme, *Developing capacity, conserving biodiversity, sustaining livelihood* (November 2004), available at: <http://www.undp.org/gef/undp-gef_publications/publications/biodiversity_ brochure2004.pdf>.

127. *See* Daniel M. Bodansky, *International law and the protection of biological diversity*, 28 VAND. J. TRANSNAT'L L. 623, 632–34 (1995). *See generally* Timothy M. Swanson, *Economics of a biodiversity convention*, 21 AMBIO 250 (1992); Jon H. Goldstein, *The prospects for using market incentives to conserve biological diversity*, 21 ENVT'L L. 985 (1991).

128. CBD, Art. 18.

129. CBD, Art. 18.3.

130. CBD, Art. 19.2.

131. CBD, Art. 16.3.

132. Philippe Sands, PRINCIPLES OF INTERNATIONAL ENVIRONMENTAL LAW I: FRAMEWORKS, STANDARDS AND IMPLEMENTATION, 385–86 (Manchester University Press, 1995) (discusses the "dismay" of the United States and other developed countries to the provisions of the CBD).

133. *See* Convention on Biological Diversity, *Parties to the Convention on Biological Diversity/Cartagena Protocol on Biosafety,* available at: <http:// www.biodiv.org/world/parties.asp>.

134. *See, e.g.,* CBD Art. 5–11, 14(1), and 20(1).

135. *Id.* Art. 20(4).

136. *See* Tinker, *supra* note 52, at 802.

137. Benjamin J. Richardson, *Environmental law in postcolonial societies: Straddling the local-global institutional spectrum*, 11 COLO. J. INT'L ENVT'L L. & POL'Y 1, 66 (2000).

138. *Id.*

139. *See generally* Amanda Hubbard, *Comment: The Convention on Biological Diversity's fifth anniversary: A general overview of the convention—Where has it been, and where is it going?*, 10 TUL. ENVT'L L.J. 415 (1997) (outlines the spotty results the CBD yielded during its first five years).

140. CBD, Art. 1.

141. CBD, Art. 16.3.

142. *See* Sands, *supra* note 132, at 385–86.

143. CBD, Art. 3 and 4. Article 4 differentiates between "components of biodiversity" and "activities and processes." Each state has exclusive jurisdiction within the limits of its national jurisdiction for "components," while jurisdiction encompasses the area both within the limits of territorial jurisdiction and beyond for "activities and processes" (Art. 4).

144. CBD, Art. 5.

145. CBD, Art. 22(2). This applies to both customary and conventional marine law.

146. *See* Anton, *supra* note 50, at 357.

147. CBD, Art. 8(a).

148. *See* Anton, *supra* note 50, at 359. Additionally, even if possible, the CBD fails to provide incentives to states to protect marine biodiversity beyond the continental shelf and EEZ (356).

149. *See Report of the Second Meeting of the COP to the Convention on Biological Diversity*, U.N. Doc. UNEP/CBD/COP/2/19, 12 (November 30, 1995) (hereinafter *Second Meeting Report*). Decision II/10 is referred to in the Ministerial Statement adopted as the "Jakarta Mandate."

150. *Id.* at 14.

151. Convention Concerning the Protection of the World Cultural and Natural Heritage, November 23, 1972, 27 U.S.T. 37, 11 I.L.M. 1358, 1037 U.N.T.S. 151 (entered into force December 17, 1975). *See* <http://www.unesco.org/whc/nwhc/pages/doc/main.htm> (hereinafter WHC).

152. WHC, Art. 1.

153. For a current list of all states parties to the WHC, including date of ratification, accession, or succession, *see* UNESCO World Heritage Centre, *States parties*, available at: <http://whc.unesco.org/wldrat.htm>.

154. Kunich, *supra* note 76, at 36–39.

155. WHC, preamble.

156. As defined in WHC, Art. 2.

157. As defined in WHC, Art. 1. Article 1 provides for three types of cultural resources: (1) monuments, which are defined as "architectural works, works of monumental sculpture and painting, elements or structures of an archaeological nature, inscriptions, cave dwellings and combinations of features, which are of outstanding universal value from the point of view of history, art or science"; (2) groups of buildings, defined as "groups of separate or connected buildings which, because of their architecture, their

homogeneity or their place in the landscape, are of outstanding universal value from the point of view of history, art or science"; and (3) sites, which are "works of man or the combined works of nature and man, and areas including archaeological sites which are of outstanding universal value from the historical, aesthetic, ethnological or anthropological point of view."

158. WHC, Art. 2.

159. WHC, Art. 4.

160. *Id.*

161. *Id.*

162. WHC, Art. 5a–e.

163. WHC, Art. 6.1.

164. WHC, Art. 6.2.

165. WHC, Art. 6.3.

166. WHC, Art. 7.

167. *See* Daniel L. Gebert, *Sovereignty under the World Heritage Convention: A questionable basis for limiting federal land designation pursuant to international agreements*, 7 S. CAL. INTERDIS. L.J. 427, 436–38 (1998).

168. WHC, Art. 3.

169. WHC, Art. 11.3.

170. *See Operational guidelines for the implementation of the World Heritage Convention*, paragraphs 6, 44(a)(vi), available at: <http://www.unesco.org/whc/opgutoc.htm>.

171. Provided for in WHC, Art. 8.

172. WHC, Art. 14.2. The IUCN was initially called the International Union for Conservation of Nature and Natural Resources. For an example of the type of detailed scientific assessment that serves as a predicate to consideration of natural sites for World Heritage listing, *see* Steven L. Chown, Ana S.L. Rodrigues, Niek J.M. Gremmen, and Kevin J. Gaston, *World heritage status and conservation of southern ocean islands,* 15 CONSERVATION BIOLOGY 550–57 (June 2001).

173. For a complete list of sites on the World Heritage List, *see* UNESCO World Heritage Centre, *World Heritage List*, available at: <http://whc.unesco.org/en/list/>.

174. *See* UNESCO World Heritage Centre, *World Heritage List*, available at: <http://whc.unesco.org/en/list/>.

175. *See* UNESCO World Heritage Centre, *Coiba National Park and its special zone of marine protection*, available at: <http://whc.unesco.org/en/list/1138>. The UNESCO website describes this situation as follows:

Coiba National Park, off the southwest coast of Panama, protects Coiba
Island, 38 smaller islands and the surrounding marine areas within the Gulf
of Chiriqui. Protected from the cold winds and effects of El Niño, Coiba's
Pacific tropical moist forest maintains exceptionally high levels of endemism
of mammals, birds and plants due to the ongoing evolution of new species. It
is also the last refuge for a number of threatened animals such as the crested
eagle. The property is an outstanding natural laboratory for scientific research
and provides a key ecological link to the Tropical Eastern Pacific for the
transit and survival of pelagic fish and marine mammals.

See also IUCN News, *Eight new world heritage sites designated*, available
at: <http://www.iucn.org/themes/wcpa/newsbulletins/news/pressreleases/
new_wh_sites_designated.pdf>.
176. *See* UNESCO World Heritage Centre, *Map of world heritage properties*, available at: <http://whc.unesco.org/en/map/> for an interactive
online map of World Heritage sites.
177. WHC, Art. 15.3.
178. *Id.*
179. WHC, Art. 17.
180. WHC, Art. 15.4.
181. *Id.*
182. WHC, Art. 16.1.
183. WHC, Art. 16.4.
184. WHC, Art. 16.5.
185. WHC, Art. 22.
186. WHC, Art. 22 a–f.
187. WHC, Art. 24.
188. WHC, Art. 25.
189. *Id.*
190. WHC, Art. 13.4.
191. WHC, Art. 11.4.
192. *Id.*
193. *Id.*
194. *See* UNESCO World Heritage Centre, *World heritage in danger list,*
available at: <http://whc.unesco.org/en/danger/>.
195. *Id.*
196. *Id.*
197. *See* Ben Boer, *World heritage disputes in Australia*, 7 J. ENVTL. L. &
LITIG. 247, 258–76 (1992) (describes several disputes rising out of World
Heritage listing proposals in Australia).

198. *See* Matthew Machado, *Mounting opposition to biosphere reserves and world heritage sites in the United States sparked by claims of interference with national sovereignty*, COLO. J. INT'L ENVTL. L. Y.B. 120 (1997); Gebert, *supra* note 167, at 427–29.

199. WHC, Art. 27.1–2. Similarly, Article 28 requires nations that receive international assistance for a World Heritage site to "take appropriate measures to make known the importance of the property for which assistance has been received and the role played by such assistance."

200. *See* Simon Lyster, INTERNATIONAL WILDLIFE LAW: AN ANALYSIS OF INTERNATIONAL TREATIES CONCERNED WITH THE CONSERVATION OF WILDLIFE, 301–2 (Cambridge University Press, 1985) (criticizes the WHC as having "proved relatively ineffectual" because, inter alia, it failed to establish "a system of administration to monitor and oversee" enforcement).

201. WHC, Art. 29.1 (provides that upon the request of a specified U.N. committee, a party "shall . . . give information on the legislative and administrative provisions which they have adopted and other action which they have taken").

202. *See* Edith Brown Weiss, *The five international treaties: A living history*, 104, in ENGAGING COUNTRIES: STRENGTHENING COMPLIANCE WITH INTERNATIONAL ENVIRONMENTAL ACCORDS (Edith Brown Weiss and Harold K. Jacobsen, eds., MIT Press, 1998).

203. *See* Brad L. Bacon, *Enforcement mechanisms in international wildlife agreements and the United States: Wading through the murk*, 12 GEO. INT'L ENVTL. L. REV. 331, 354–55 (1999).

204. Weiss, *supra* note 202, at 93–105, 125–35.

205. WHC, Art. 2.

206. While the WHC itself only mentions the listing of sites, the Operational Guidelines provide that parties may delist a site if a host country fails to protect it. *See Operational guidelines for the implementation of the World Heritage Convention*, available at: <http://whc.unesco.org/en/guidelines>.

207. John Charles Kunich, *World heritage in danger in the hotspots*, 78 IND. L. J. 619, 646–56 (2003).

208. A list of the current parties to the London Dumping Convention can be found at Office for the London Convention, *Parties to the London Convention as of June, 2005*, available at: <http://www.londonconvention.org/PartiesToLC .htm>; <http://www.londonconvention.org/main.htm>. Closely related to the London Convention is the 1973 International Convention for the Prevention of Pollution from Ships and the 1978 protocol thereto, often known

collectively as MARPOL 73/78 (an abbreviation of "marine pollution"). *See* International Maritime Organization, *International Convention for the Prevention of Pollution from Ships, 1973, as modified by the Protocol of 1978 relating thereto (MARPOL 73/78)*, available at: <http://www.imo.org/Conventions/contents .asp?doc_id=678&topic_id=258>. MARPOL 73/78 focuses on proactive prevention of operational discharges (such as oil spills), rather than the intentional dumping of wastes covered by the London Convention. Although MARPOL 73/78 is credited with some reduction of oil tanker pollution, many ports still have inadequate reception facilities for oil-containing wastes, and enforcement has been less than vigorous. *See* Andrew Griffin, *Marpol 73/78 and vessel pollution: A glass half full or half empty?*, 1 IND. J. GLOBAL LEGAL STUD. 489, 505 (1994).

209. Convention on the Prevention of Marine Pollution by Dumping of Waste and Other Matters (London Dumping Convention), 1972, Annex I.

210. London Dumping Convention, Annex II.

211. London Dumping Convention, 1993 Amendments.

212. 1996 Protocol on the Prevention of Marine Pollution by Dumping of Waste and Other Matters, Art. 5. Incineration of wastes at sea is also regulated under the MPRSA.

213. London Dumping Convention, Art. 7.

214. *Id.*

215. *See* International Maritime Organization, available at: <http://www.imo.org/home.asp>.

216. Blacklisted substances include high-level radioactive wastes, mercury, cadmium, persistent plastics, and other highly toxic or long-lived materials. *See Convention on the Prevention of Marine Pollution by Dumping of Wastes and Other Matter*, Annexes I, II, and III, available at: <http://inter national.nos.noaa.gov/conv/london.html#ANNEXES>.

217. Cruise ships are a particularly notorious example of deliberate despoiling of the marine environment through massive dumping of wastes. These ships are often immense vessels, carrying thousands of passengers and crew members on each voyage, and they generate and dispose of titanic quantities of waste at sea or in poorly equipped shore facilities. *See* GESAMP (IMO/FAO/UNESCO/WMO/IAEA/UN/UNEP, Joint Group of Experts on the Scientific Aspects of Marine Environmental Protection), and Advisory Committee on Protection of the Sea, *A sea of troubles*, REP. STUD. GESAMP no. 70, 24 (2001), available at: <http://gesamp.imo.org/no70/ index.htm>.

218. *Id.*

219. *See* The Fridtjof Nansen Institute, *1996 Protocol to the Convention on the Prevention of Marine Pollution by Dumping of Wastes and Other Matter, 1972 (1996 Protocol to the London Convention 1972)*, YEARBOOK OF INTERNATIONAL CO-OPERATION ON ENVIRONMENT AND DEVELOPMENT, available at: <http://www.greenyearbook.org/agree/mar-env/m-london2.htm>.

220. *See 1996 Protocol to the Convention on the Prevention of Marine Pollution by Dumping of Wastes and Other Matter, 1972 and resolutions adopted by the special meeting*, Annexes 1–3, available at: <http://international.nos.noaa.gov/conv/lonprot.html#ANNEXES>.

221. *See* Office for the London Convention, *London Convention 1972*, available at: <www.londonconvention.org>. Additionally, 44 percent of ocean pollution is caused by runoff and land-based discharges, 33 percent by land-based discharges through the atmosphere, and 12 percent by maritime transportation activities.

222. *Id.* Between 1992 and 1995 discharges of industrial waste averaged between 4.5 and 6 million tons annually. Discharges of sewage sludge actually increased during the period 1992 to 1994 from 12.5 to 16.5 million tons. Discharges of dredged material range between 250 to 650 million tons annually, inclusive of internal waters.

223. *See* Olav Schram Stokke, *Beyond dumping? The effectiveness of the London Dumping Convention*, 40, in YEARBOOK OF INTERNATIONAL CO-OPERATION ON ENVIRONMENT AND DEVELOPMENT 1998/1999 (Olav Schram Stokke, ed., Earthscan Publications, 1998/1999).

224. London Dumping Convention, Art. 15.

225. *See* Stokke, *supra* note 223 (reports that, on average, half of the member states did not submit the required reports on ocean dumping in areas under their control).

226. *See id.* at 44–45.

227. *See id.* at 39, 46.

228. Convention on International Trade in Endangered Species of Wild Flora and Fauna, March 3, 1973, 993 U.N.T.S. 243 (1976) (entered into force July 1, 1975) (hereinafter CITES).

229. A list of parties to CITES in chronological order can be found at CITES, *List of contracting parties (in order of entry into force)*, available at: <http://www.cites.org/eng/disc/parties/chronolo.shtml>. There is also an alphabetical list at CITES, *List of contracting parties in alphabetical order*, available at: <http://www.cites.org/eng/disc/parties/alphabet.shtml>.

230. *See* CITES, *What is CITES?*, available at: <http://www.cites.org/eng/disc/what.shtml>.

231. Endangered Species Act (ESA), 16 U.S.C. 1531–1544, available at: <http://www4.law.cornell.edu/uscode/html/uscode16/usc_sup_01_16_10_35.html>.

232. CITES, Arts. III–V.

233. Appendix I lists species "threatened with extinction." It includes all apes, lemurs, the giant panda, many South American monkeys, the great whales, cheetahs, leopards, tigers, Asian and African elephants, all rhinoceroses, any birds of prey, cranes and pheasants, all sea turtles, some crocodile and lizards, giant salamanders, some mussels, orchids, and cacti. Appendix II lists species and specimens that are not yet threatened with extinction but which "may become" so if trade in them is not controlled. Appendix II includes primates, cats, otters, smaller whales, dolphins and porpoises, some birds of prey, tortoises, crocodiles, fur seals, the black stork, birds of paradise, the coelacanth, some snails, birdwing butterflies, and black coral. Appendix III contains species listed by nations that have stricter legislation than the CITES requirements, restricting import and export of species not listed in Appendices I or II. Nations can list such species in Appendix III, after which the other parties must regulate trade in those species. *See* Patricia Birnie, *The case of the Convention on Trade in Endangered Species*, 233, in ENFORCING ENVIRONMENTAL STANDARDS: ECONOMIC MECHANISMS AS VIABLE MEANS? (Jochen Abr. Frowein, et al., eds., 1996); Michelle Ann Peters, *The Convention on International Trade in Endangered Species: An answer to the call of the wild?* 10 CONN. J. INT'L L. 169, 176 (1994).

234. Margaret Rosso Grossman, *Habitat and species conservation in the European Union and the United States*, 45 DRAKE L. REV. 19, 20–21 (1997).

235. CITES, Art. VIII(1)(a), 1(b), and (2). Under this Article, all parties "shall" take appropriate measures to enforce the provisions of CITES and to prohibit trade in specimens taken in violation thereof. These shall include measures to penalize trade in or possession of (or both) such specimens, and measures to provide for the confiscation or return to the state of export of such specimens. The measures mentioned in Article VIII(1) must include, inter alia, appropriate penalties for trading in prohibited specimens and the power to confiscate species and provide for their return to the state of export. Failure to enact penalties constitutes a violation of CITES. The provisions of Article VIII indicate that each party has the obligation to implement CITES through its own domestic legislation, as the United States has done with the ESA.

236. *Id.*, Art. VIII(7). The report is to consist of information on the number and type of permits allocated with respect to species protected by CITES.

237. *Id.*, Art. XIII(1).

238. *Id.*, Art. XVIII(1) and (2). There is a provision for voluntary referral of a dispute to arbitration, with specific mention of the Permanent Court of Arbitration at The Hague.

239. *See* Carlo A. Balistrieri, *CITES: The ESA and International Trade*, 8 NAT. RESOURCES & ENVT'L L. 33, 54 (1993).

240. *See, e.g.*, Michael Glennon, *Has international law failed the elephant?*, 84 AM. J. INT'L L. 1, 20 (1990) (states that it "seems fair to conclude that throughout the 1980s, the trade in elephant body parts, including ivory, boomed despite the CITES protective regime for a fairly obvious reason: CITES did not sufficiently diminish the incentives of producers, middlemen or consumers"); Julie Cheung, *Implementation and enforcement of CITES: An assessment of tiger and rhinoceros conservation policy in Asia*, 5 PAC. RIM L. & POL'Y J. 125–26 (1995) (questions whether CITES has been successful in protecting tigers and rhinos); Joonmoo Lee, *Poachers, tigers and bears . . . Oh my! Asia's illegal wildlife trade*, 16 NW. J. INT'L L. & BUS. 497, 503–4 (1996) (critiques CITES as "being largely powerless to enforce its resolutions" while noting some success stories as well); Kevin D. Hill, *The Convention on International Trade in Endangered Species: Fifteen years later*, 13 LOY. L.A. INT'L & COMP. L.J. 231, 277 (1990) (notes that some developing countries do not want to shut down the international trade in endangered species because to do so would eliminate an important source of income for their economies).

241. CITES, Art. XV(3), Art. XVI(2).

242. Sands, *supra* note 132, at 378–79.

243. *See* John C. Kunich, *The fallacy of deathbed conservation under the Endangered Species Act*, 24 ENVT'L L. 501, 5022–28 (1994).

244. There are numerous other international agreements aimed at the preservation of specific types of living things, whether plants, birds, marine species, or particular subsets thereof. These can be helpful within their limited ambit, but generally suffer from the same problems inherent in the ESA approach, and are not appropriate for broader hotspots preservation. *See, e.g.*, 1950 International Convention for the Protection of Birds, October 18, 1950, 638 U.N.T.S. 185 (entered into force January 17, 1963); 1951 International Convention for the Establishment of the European and Mediterranean Plant Protection Organization, April 18, 1951, U.K.T.S. 44 (1956) (entered into force November 1, 1953).

CHAPTER THREE

1. *See* Pew Oceans Commission, *America's living oceans: Charting a course for sea change*, 27 (May 2003), available at: <http://www.oceanconservancy.org/site/DocServer/oceans_report.pdf?docID=242> (hereinafter *Sea change*). The United States has the largest EEZ in the world, spanning an area of 4.5 million square miles, 23 percent larger than the nation's land area (31).

2. 33 U.S.C. 1401–1445, available at: <http://www4.law.cornell.edu/uscode/html/uscode33/usc_sup_01_33_10_27.html>.

3. 33 U.S.C. 1411.

4. 33 U.S.C. 1412.

5. *See* U.S.C. 1412(a).

6. *See* U.S.C. 1413(b).

7. *See* <http://www.ncseonline.org/nle/crsreports/briefingbooks/laws/f.cfm>.

8. 33 U.S.C. 1415.

9. 33 U.S.C. 1433(a).

10. 33 U.S.C. 1433(b)(1)(A).

11. 33 U.S.C. 1434.

12. 33 U.S.C. 1431.

13. 33 U.S.C. 1431(b)(2).

14. 33 U.S.C. 1434(d)(1)(A).

15. 33 U.S.C. 1434(d).

16. 16 U.S.C. 1801–1883. The Magnuson-Stevens Fishery Conservation and Management Act, originally enacted in 1976, provides for the conservation and management of fishery resources within the United States' 200-mile EEZ. The act does not significantly diminish the powers of the states to regulate fishing in the waters off their respective coasts, but, thanks to 1996 amendments in the form of the Sustainable Fisheries Act, does empower the federal government to identify, respond to, and regulate overfishing. In this regard, fishery plans are to consider "essential fish habitat" among other criteria weighed in evaluating these plans.

17. 33 U.S.C. 1434(a)(5).

18. *See, e.g.,* Craft v. National Park Service, 34 F.3d 918 (9th Cir. 1994); United States v. Fisher, 22 F.3d 262 (11th Cir. 1994); and Personal Watercraft Industry Association v. Dep't of Commerce, 48 F.3d 540 (D.C. Cir. 1995). In each of these cases, restrictions were upheld on private recreational activities such as diving, small craft use, and salvage collecting.

19. *See* National Marine Sanctuaries website, available at: <http://sanctuaries.noaa.gov>. MPAs are located far and wide, including along the Atlantic and Pacific coasts, near Hawaii, in the Gulf of Mexico, and in other U.S. territories. A total of approximately 150,000 square miles falls within these 14 MPAs, in the aggregate.

20. 65 Fed. Reg. 34909 (May 31, 2000).

21. *See* U.S. Department of Commerce/NOAA, *Marine protected areas of the United States*, available at: <http://mpa.gov> (shows a list of MPAs grouped by the legal authority used to establish them).

22. U.S. Department of Commerce/NOAA, *Executive Order 13158 of May 26, 2000*, available at: <http://mpa.gov/executive_order/execordermpa.pdf>.

23. Daniel Pauly, et al., *Towards sustainability in world fisheries*, 418 NATURE 689, 692–94 (2002), available at: <http://courses.washington.edu/susfish/speakers/punt1.pdf>.

24. *See, e.g.,* Callum M. Roberts, et al., *Effects of marine preserves on adjacent fisheries*, 294 SCIENCE 1920 (2001); Callum M. Roberts, et al., *Marine biodiversity hotspots and conservation priorities for tropical reefs*, 295 SCIENCE 1280 (2002).

25. Coastal Zone Management Act of 1972, 16 U.S.C. 1451–1465, available at: <http://www.access.gpo.gov/uscode/title16/chapter33_.html>.

26. 16 U.S.C. 1451(a–c).

27. 16 U.S.C. 1463(b).

28. 16 U.S.C. 1454–1456.

29. 16 U.S.C. 1456(c).

30. 16 U.S.C. 1456(c)(1)(A) reads as follows:

Each federal agency activity within or outside the coastal zone that affects any land or water use or natural resource of the coastal zone shall be carried out in a manner which is consistent to the maximum extent practicable with the enforceable policies of approved state management programs.

31. 16 U.S.C. 1451(I).

32. *See* NOAA Office of Ocean and Coastal Resource Management, *Celebrating 30 years of the Coastal Zone Management Act*, available at: <http://www.ocrm.nos.noaa.gov/czm/> for a listing and description of the various state management programs under CZMA.

33. Endangered Species Act (ESA), 16 U.S.C. 1531–1544, available at: <http://www4.law.cornell.edu/uscode/html/uscode16/usc_sup_01_16_10_35.html>.

34. 16 U.S.C. 1532(19).

35. Either the Department of Interior for terrestrial and freshwater species, or the Department of Commerce for marine species. *See* 16 U.S.C. 1536(a)(1–2).

36. 16 U.S.C. 1536(a)(1–2).

37. *See, e.g.,* John Copeland Nagle, *Playing Noah,* 82 MINN. L. REV. 1171 (1988).

38. For example, twelve salmon and steelhead trout runs are listed as critical habitat. Yet, despite the expenditure of $3.5 billion dollars since 1980 on restoring these runs, these runs have continued to decline and wild salmon have nearly vanished from the Columbia River Basin. This failure has been blamed on the fragmentation of responsibility for planning, implementing, and funding the protection efforts, along with the failure to establish firm restoration goals and the lack of legal and institutional mechanisms to ensure that the goals are achieved. *See Sea change, supra* note 1, at 28.

39. Marine Mammal Protection Act, 16 U.S.C. 1361–1407, available at: <http://www4.law.cornell.edu/uscode/html/uscode16/usc_sup_01_16_10_31_20_I.html>. Both the ESA and the MMPA are discussed in U.S. Commission on Ocean Policy, *An ocean blueprint for the 21st century, Final Report, Washington, D.C.,* Ch. 20 (2004), available at: <http://www.ocean commission.gov/documents/full_color_rpt/000_ocean_full_report.pdf> (hereinafter *Ocean blueprint*).

40. 16 U.S.C. 1373–1374.

41. 16 U.S.C. 1401–1402.

42. 16 U.S.C. 1361(6).

43. 16 U.S.C. 1362(8).

44. 16 U.S.C. 1362(7), 1371(a).

45. 16 U.S.C. 1362(1)(A–C).

46. 16 U.S.C. 1371(a)(2).

47. *See* U.S. Environmental Protection Agency, *U.S. coral reef task force,* available at: <http://www.epa.gov/OWOW/oceans/coral/taskforce.html>.

48. *See generally* Robin Kundis Craig, *Taking the long view of ocean ecosystems: Historical science, marine restoration, and the Oceans Act of 2000,* 29 ECOLOGY L.Q. 649 (2002); *Ocean blueprint, supra* note 39, at Ch. 21 (focuses on coral reefs as key marine habitats).

49. *See* National Oceans Office, *Australia's oceans policy,* available at: <http://www.oceans.gov.au/content_policy_v1/default.jsp>.

50. *See Id.*; Tundi Agardy, *Global Trends in Marine Protected Areas,* 52–53, available at: <http://64.233.187.104/search?q=cache:ru4D9HVtifkJ:www .oceanservice.noaa.gov/websites/retiredsites/natdia_pdf/8agardy.pdf+Tundi +Agardy,+Global+Trends+in+Marine+Protected+Areas&hl=en>.

51. *See* National Oceans Office, *Assessment report, impacts: Identifying disturbances, the South-East Regional Marine Plan*, available at: <http://www.oceans.gov.au/pdf/identifying_disturbances.pdf>.

52. *See* National Oceans Office, *Assessment report, ocean management—the legal framework, the South-East Regional Marine Plan*, available at: <http://www.oceans.gov.au/pdf/legal_framework.pdf>.

53. Oceans and Law of the Sea, Division for Ocean Affairs and the Law of the Sea, *Continental Shelf (Living Natural Resources) Act 1968–1973*, available at: <http://www.un.org/Depts/los/LEGISLATIONANDTREATIES/PDFFI LES/AUS_1973_CS.pdf>.

54. Oceans and Law of the Sea, Division for Ocean Affairs and the Law of the Sea, *Seas and Submerged Lands Act 1973, as amended by the Maritime Legislation Amendment Act 1994*, available at: <http://www.un.org/Depts/los/LEGIS LATIONANDTREATIES/PDFFILES/aus_1994_sea_act.pdf>.

55. *See* Australian Government Attorney-General's Department, *Environment Protection and Biodiversity Conservation Act 1999*, available at: <http://scaleplus.law.gov.au/html/pasteact/3/3295/top.htm>; Australian Government Department of the Environment and Heritage, *About the EPBC Act*, available at: <http://www.ea.gov.au/epbc/about/index. html>; Australian Government Attorney-General's Department, *Environment Protection and Biodiversity Conservation Regulations 2000—list of regulations*, available at: <http://scaleplus.law.gov.au/html/pastereg/3/1619/top .htm>.

56. National Archives of Australia, *Seas and Submerged Lands Acts 1973*, available at: <http://www.foundingdocs.gov.au/item.asp?dID=30>.

57. National Archives of Australia, *Coastal Waters (State Powers) Act 1980*, available at: <http://www.foundingdocs.gov.au/item.asp?dID=31>.

58. Australian Government Attorney-General's Department, *Navigation Act 1912*, available at: <http://scaleplus.law.gov.au/html/pasteact/1/516/ top.htm>.

59. Australian Government Attorney-General's Department, *Protection of the Sea (Prevention of Pollution from Ships) Act 1983*, available at: <http:// scaleplus.law.gov.au/html/pasteact/0/221/top.htm>.

60. Australian Government Attorney-General's Department, *Environment Protection (Sea Dumping) Act 1981*, available at: <http://scaleplus.law. gov.au/html/pasteact/0/260/top.htm>.

61. The Biological Diversity Advisory Committee has a website available at: <http://www.deh.gov.au/biodiversity/science/bdac/>.

62. *See* Australian Government Department of the Environment and Heritage, *State of the Marine Environment Report for Australia: Pollution—Technical Annex 2*, available at: <http://www.deh.gov.au/coasts/publicatio ns/somer/annex2/>.

63. *See* Great Barrier Reef Marine Park Authority, The World Bank, and The World Conservation Union (Graeme Kelleher, Chris Bleakley, and Sue Wells, eds.), A GLOBAL REPRESENTATIVE SYSTEM OF MARINE PROTECTED AREAS, VOL. I, 154–99 (The World Bank, Washington, D.C., 1995) (hereinafter GLOBAL SYSTEM).

64. This has now become the Natural Resource Management Ministerial Council.

65. *See* Australian Government Department of the Environment and Heritage, *National Representative System of Marine Protected Areas (NRSMPA)*, available at: <http://www.deh.gov.au/coasts/mpa/nrsmpa/>.

66. National Oceans Office, available at: <http://www.oceans.gov.au/ pdf/legal_framework.pdf> at 50. The Strategic Plan can be found at Australian Government Department of the Environment and Heritage, *Strategic plan of action for the National Representative System of Marine Protected Areas: A guide for action by Australian governments (ANZECC Task Force on Marine Protected Areas)*, available at: <http://ea.gov.au/coasts/mpa/nrsmpa/spa.html>.

67. *Id.*

68. Great Barrier Reef Marine Park Act of 1975, available at: <http:// scaleplus.law.gov.au/html/pasteact/0/306/top.htm>.

69. *Great Barrier Reef Marine Park (aquaculture) regulations of 2000*, available at: <http://scaleplus.law.gov.au/html/pastereg/3/1578/top.htm>.

70. *Great Barrier Reef region (prohibition of mining) regulations of 1999*, available at: <http://scaleplus.law.gov.au/html/pastereg/3/1573/top.htm>.

71. *See* Australian Government, Great Barrier Reef Marine Park Authority, *The Great Barrier Reef Marine Park*, available at: <http://www.gbrmpa .gov.au/>.

72. Australia has a total of four marine WHC sites. In addition to the Great Barrier Reef, these are Macquarie Island, Lord Howe Island, and the Heard and McDonald islands.

73. New Zealand Biodiversity Strategy website, available at: <http:// www.biodiversity.govt.nz/picture/doing/nzbs/index.html>.

74. *See* New Zealand Biodiversity Strategy, *New Zealand's marine reserves*, available at: <http://www.biodiversity.govt.nz/seas/biodiversity/protected/ reserves.html>.

75. *See* New Zealand Biodiversity Strategy, *Marine protected areas,* available at: <http://www.biodiversity.govt.nz/seas/biodiversity/protected/index.html>.

76. *See Marine protected areas,* available at: <http://www.biology.duke.edu/bio217/2002/fish/mpa.html>.

77. *See* Lauretta Burke, Elizabeth Selig, and Mark Spalding, REEFS AT RISK IN SOUTHEAST ASIA, 20–32 (World Resources Institute, 2002) (hereinafter REEFS AT RISK).

78. *Id.* at 8.

79. *Id.* at 13–16.

80. *Id.* at 36–38.

81. *Id.* at 8. Vitally important coral reefs in Cambodia, Singapore, Taiwan, the Philippines, Vietnam, China, the Spratly Islands, Malaysia, and Indonesia are included in this list of reefs at severe risk.

82. *Id.* at 28–29, 33–52. *See* GLOBAL SYSTEM, *supra* note 63, Vol. III, at 107–36.

83. UP-MSI, ABC, ARCBC, DENR, ASEAN, MARINE PROTECTED AREAS IN SOUTHEAST ASIA (ASEAN Regional Centre for Biodiversity Conservation, Department of Environment and Natural Resources, 2002), available at: <http://www.arcbc.org.ph/MarinePA/abstract.htm>.

84. *Id.* at 28.

85. *Id.* at 37–38.

86. *Id.* at 41.

87. *Id.* at 52.

88. *Id.*

89. *Id.* at 62.

90. REEFS AT RISK, *supra* note 77, at 28.

91. *Id.*

92. *Id.*

93. *See* Mary Gray Davidson, *Protecting coral reefs: The principal national and international legal instruments,* 26 HARV. ENVTL. L. REV. 499, 504–8, 510–19, 526–40 (2002) (describes the various national and international laws applicable to coral reef protection and the laws' failure to provide effective safeguards for many important reefs worldwide).

CHAPTER FOUR

1. John Charles Kunich, ARK OF THE BROKEN COVENANT: PROTECTING THE WORLD'S BIODIVERSITY HOTSPOTS, 13–18 (Praeger, 2003).

2. The computer-animated film *Finding Nemo* (Disney/Pixar, 2003) features a blue fish named Dory, for which comedian/actress Ellen De-Generes provides the voice. Dory suffers from short-term memory loss, which causes her, inter alia, to refer to the young clownfish named Nemo by such misnomers as Harpo, Fabio, and Elmo.

3. Quincy Jones, quoted in the documentary film *Listen up: The lives of Quincy Jones* (Warner Bros., 1990). *See also* John Charles Kunich, *Losing Nemo: The mass extinction now threatening the world's ocean hotspots,* 30 CO-LUMBIA JOURNAL OF ENVIRONMENTAL LAW 1–133 (2005).

4. *See* GESAMP (IMO/FAO/UNESCO/WMO/IAEA/UN/UNEP, Joint Group of Experts on the Scientific Aspects of Marine Environmental Protection), and Advisory Committee on Protection of the Sea, *A sea of troubles.* REP. STUD. GESAMP no. 70, 26 (2001), available at: <http://gesamp.imo.org/no70/index.htm> (hereinafter *Sea of troubles*) (cites adoption of an action plan addressing pollution, sustainable use of resources, and effective management of coastal areas in the Mediterranean).

5. Despite the Chicago Cubs' remarkably and unexpectedly successful season in 2003, which saw them improve by twenty-one victories over the previous year, win the National League Central Division title, and defeat the heavily favored Atlanta Braves in the Division Series, they fell one victory short of their first trip to the World Series since 1945. The Cubs have not won a World Series since 1908. *See* Mike Riley, *Cubs reeled in again,* CHI-CAGO SUN-TIMES (October 16, 2003).

6. *See generally* International Court of Justice website, available at: <http://www.icj-cij.org/icjwww/icjhome.htm>.

7. *See generally* International Criminal Court website, available at: <http://www.icc-cpi.int/home.html&l=en>.

8. In physics and other fields, a ''thought experiment'' (from the German *Gedankenexperiment*) is an attempt to solve a problem using the power of human imagination. Such methods are particularly useful in situations where an actual empirical experiment is either impossible or impractical.

9. *See Sea of troubles, supra* note 4, at 28–29 (lists the causes of failure to protect our oceans, including economic constraints; low priority given to environmental protection; weakness of national structures; deficiencies in policies and practices; insufficient public awareness; ineffective communication between scientists and government policy-makers; and the broad, fragmented approach of international law that is not translated into specific, well-defined actions and priorities).

10. *See generally* Mary Renault, THE NATURE OF ALEXANDER (Pantheon, 1975).

11. *See* J.K. Pinnegar, et al., *Trophic cascades in benthic marine ecosystems: Lessons for fisheries and protected-area management,* 27 ENVIRONMENTAL CONSERVATION 179–200 (2000), available at: <http://www.unice.fr/LEML/Pages/Pub_LEML/Pinnegar_et_al_2000.pdf>; Marten Scheffer, et al., *Catastrophic shifts in ecosystems,* 413 NATURE 591–96 (2001) (details the domino effect, or trophic cascade, that often results from over-exploitation of limited portions of a marine ecosystem, and the importance of using MPAs to guard against such disastrous chains of events).

12. *See* Gary W. Allison, Jane Lubchenco, and Mark H. Carr, *Marine reserves are necessary but not sufficient for marine conservation,* ECOLOGICAL APPLICATIONS 8(1) Supplement, S79, S81–85 (1998), available at: <http://bio.research.ucsc.edu/people/carr/publications/carr/Allison%20et%20al.%20Ecol%20App%201998.pdf>; Steven N. Murray, et al., *No-take reserve networks: Sustaining fishery populations and marine ecosystems,* 24 FISHERIES 11, 15–21 (1999).

13. *See Sea of troubles, supra* note 4, at 32–34 (summarizes many significant problem areas in marine environmental protection and proposes corrective actions).

14. *See* GESAMP (IMO/FAO/UNESCO/WMO/IAEA/UN/UNEP, Joint Group of Experts on the Scientific Aspects of Marine Environmental Protection), and Advisory Committee on Protection of the Sea, *Marine biodiversity: Patterns, threats, and conservation needs.* REP. STUD. GESAMP no. 62, 7 (1997), available at: <http://gesamp.imo.org/no62/index.htm>.

15. *Id.*

16. *See* GESAMP (IMO/FAO/UNESCO/WMO/IAEA/UN/UNEP, Joint Group of Experts on the Scientific Aspects of Marine Environmental Protection), and Advisory Committee on Protection of the Sea, *Protecting the oceans from land-based activities: Land-based sources and activities affecting the quality and uses of the marine, coastal and associated freshwater environment.* REP. STUD. GESAMP no. 71, 9–37 (2001), available at: <http://gesamp.imo.org/no71/index.htm>.

17. *Id.* at 17–20.

18. *Id.* at 21–23.

19. *Id.* at 20–27.

20. *See* International Maritime Organization, *Invasive species: The problem,* available at: <http://globallast.imo.org/problem.htm>; U.S. Commission on

Ocean Policy, *An ocean blueprint for the 21st century, Final Report, Washington, D.C.*, Ch. 17 (2004), available at: <http://www.oceancommission.gov/documents/full_color_rpt/000_ocean_full_report.pdf> (hereinafter *Ocean blueprint*); James T. Carlton, *Invasive species and biodiversity management*, 195–212, in THE SCALE AND ECOLOGICAL CONSEQUENCES OF BIOLOGICAL INVASIONS IN THE WORLD'S OCEANS (O.T. Sandlund, P.J. Schei, and A. Viken, eds., Kluwer Academic Publishers, 1999).

21. *See* National Research Council, MARINE PROTECTED AREAS: TOOLS FOR SUSTAINING OCEAN ECOSYSTEMS 71–96 (National Academies Press, 2001) (hereinafter MARINE PROTECTED AREAS).

22. *See* Mark H. Carr, et al., *Comparing marine and terrestrial ecosystems: Implications for the design of coastal marine reserves*, ECOLOGICAL APPLICATIONS 13(1) Supplement, S90, S92–93, S95–103 (2003), available at: <http://bio.research.ucsc.edu/people/carr/publications/carr/carr-et-al-2003_ecological%20applications.pdf>.

23. *See* MARINE PROTECTED AREAS, *supra* note 21, at 97–111.

24. *Id.* at 97–98. Of course, a purely hotspots-based approach would not be favored by those scientists who see greater benefits offered by one of the other methods of assigning conservation priorities, such as Global 200 Ecoregions or WORLDMAP. *See* Kunich, *supra* note 1, at 36–39.

25. *Id.* at 111–18.

26. *Id.* at 118–23. *See* Mark H. Carr, *Marine protected areas: Challenges and opportunities for understanding and conserving coastal marine ecosystems*, 27 ENVIRONMENTAL CONSERVATION (2), 106–9 (2000), available at: <http://bio.research.ucsc.edu/people/carr/publications/carr/Carr%20Envir%20Cons%202000.pdf>.

27. *See* Pew Oceans Commission, *Marine reserves: A tool for ecosystem management and conservation*, 1–2 (2002), available at: <http://www.pewoceans.org/reports/pew_marine_reserves.pdf> (hereinafter *Marine reserves*).

28. *Id.* at 2. Such areas are often called "fully protected marine reserves."

29. *See* Graeme Kelleher, *Guidelines for marine protected areas*, IUCN, 51–52, 89–96 (1999), available at: <www.iucn.org/themes/wcpa/pubs/pdfs/mpa_guidelines.pdf> (hereinafter *Guidelines*).

30. *Marine reserves*, *supra* note 27, at 2.

31. *Id.* at 14–21 (summarizes the main threats to marine biodiversity, including overfishing, habitat alteration, by-catch, recreational threats, pollutants, runoff from land, introduced/invasive species, aquaculture, climate change, and coastal development).

32. MARINE PROTECTED AREAS, *supra* note 21, at 126–44.

33. *See* Ussif R. Sumaila, et al., *Addressing ecosystem effects of fishing using marine protected areas*, 57 ICES JOURNAL OF MARINE SCIENCES 752–60 (2000), available at: <http://home.imf.au.dk/eduard/files/1Gu%E9nette 2000.pdf>; Daniel Pauly, et al., *Towards sustainability in world fisheries*, 418 NATURE 689, 690–92 (2002), available at: <http://courses.washington .edu/susfish/speakers/punt1.pdf>; Ransom A. Myers and G. Mertz, *The limits of exploitation: A precautionary approach*, 8 ECOLOGICAL APPLICA- TIONS 165–69 (1998), available at: <http://fish.dal.ca/~myers/papers/ Papers-1996–2000/limit_exploit.pdf>.

34. MARINE PROTECTED AREAS, *supra* note 21, at 123–25.

35. *See Guidelines*, *supra* note 29, at 43–50.

36. *Marine reserves*, *supra* note 27, at 31.

37. *Id.* at 35–37. MPAs should be as varied as the biodiversity and habitats they are intended to shelter, and no single size, shape, set of al- lowable activities, degree of connectivity, or other parameters can justifiably be implemented across the board.

38. *Id.* There are other lists of criteria for selecting MPAs, including one from the IUCN. *See Guidelines*, *supra* note 29, at 40–41. The IUCN proposes some criteria not on the Pew Oceans Commission list, including the degree of genetic diversity within species in the MPA; value for scientific research; accessibility for education, tourism, and recreation; social and political ac- ceptability; community support; compatibility with existing uses, especially local ones; ease of management; and the extent to which the area has been spared from human-induced change.

39. *See* Lydia K. Bergen and Mark H. Carr, *Establishing marine reserves: How can science best inform policy?*, 45 ENVIRONMENT (2), 8, 12–18 (2003), available at: <http://bio.research.ucsc.edu/people/carr/publications/carr/ Bergen%20and%20Carr%20Environment-2003.pdf>.

40. *See* the computer-animated film *Finding Nemo* (Disney/Pixar, 2003) (in which several sharks meet periodically as a support group to strengthen one another's resolve to refrain from eating their fellow fish).

41. Kunich, *supra* note 1, at 146–62 (proposes a Vital Ecosystem Pre- servation Act, or VEPA).

42. Tropical Forest Conservation Act of 1998, 22 U.S.C. 2431 (1998) (amends 22 U.S.C. 2151 [1961] [the Foreign Assistance Act of 1961]), available at: <http://caselaw.lp.findlaw.com/casecode/uscodes/22/chap ters/32/subchapters/iv/toc.html>.

43. 22 U.S.C. 2431(a)(2–7).

44. For example, to qualify for a debt-for-nature swap under this act, a nation must, inter alia, be one whose government (1) is democratically elected; (2) has not repeatedly provided support for acts of international terrorism; (3) is not failing to cooperate on international narcotics control matters; and (4) does not engage in a consistent pattern of gross violations of internationally recognized human rights (22 U.S.C. 2430b(a)(1–4)). Additionally, the nation must be either a low-income country (with a per capita income less than $725) or a middle-income country (with a per capita income more than $725 but less than $8,956) (22 U.S.C. 2431a(5)(A)(i–ii)). It must also be "a country that contains at least one tropical forest that is globally outstanding in terms of its biological diversity or represents one of the larger intact blocks of tropical forests left, on a regional, continental, or global scale" (22 U.S.C. 2431a(5)(B)). Other requirements include arrangements with various international funds, including the International Monetary Fund for adjustment loans, and formulation of financing programs with commercial bank lenders. The president has discretion to determine whether a nation meets the above standards, and thus is eligible for benefits (22 U.S.C. 2431f). The U.S. secretary of state may enter into Tropical Forest Agreements with eligible countries to operate the funds created by this act for such purposes as establishing parks and reserves, promoting sustainable use of plant and animal species, and identification of medicinal uses of tropical forest plant life; funds may also be used for training programs for scientists and support of livelihoods of individuals living in or near tropical forests to prevent exploitation of the environmental resources (22 U.S.C. 2431g).

45. Jennifer A. Loughrey, *The Tropical Forest Conservation Act of 1998: Can the United States really protect the world's resources?—The need for a binding international treaty convention on forests,* 14 EMORY INT'L L. REV. 315, 328–37 (2000) (discusses the merits and shortcomings of this act). *See also* Nancy Knupfer, *Debt-for-nature swaps: Innovation or intrusion?*, 4 N.Y. INT'L L. REV. 86, 88 (1991); Paul J. Ferraro and Randall A. Kramer, *Compensation and economic incentives: Reducing pressure on protected areas,* 187–211, in LAST STAND: PROTECTED AREAS & THE DEFENSE OF TROPICAL BIODIVERSITY (R. Kramer, et al., eds., Oxford, 1997).

46. In a debt buy-back, the debtor nation purchases its debt at a reduced price.

47. In a debt restructuring agreement, the original debt agreement is cancelled (a percentage of the face value of the debt is reduced) and a new agreement is created that provides for an annual amount of money in local currency to be deposited into a fund for conservation projects.

48. A three-party swap works as follows: An NGO (usually a conservation group) buys a hard-currency debt on the secondary market that is owed to commercial banks, or a public/official debt owed to a creditor government at a discount rate, and then renegotiates the debt obligation with the creditor nation. The money generated from the renegotiated debt, to be repaid in local currency, is usually put into a fund that can allocate grants for conservation projects.

49. Public Law 107–26. *See* Report 107–19, *Reauthorization of the Tropical Forest Conservation Act of 1998 Through Fiscal Year 2004*, 107th Cong., 1st Sess., June 28, 2001, available at: <http://lugar.senate.gov/pressapp/record .cfm?id=226907>.

50. Congressional Research Service Report for Congress, *Debt-for-Nature Initiatives and the Tropical Forest Conservation Act: Status and Implementation* (February 13, 2002), Library of Congress Order Code RL31286.

51. *Id.* at 13.

52. *See* MARINE PROTECTED AREAS, *supra* note 21, at 54–60.

53. *Id.* at 43–46.

54. *Id.* at 46–54, 60–66. *See also Marine reserves*, *supra* note 27, at 22–28 (describes the efficacy of marine reserves and the several benefits that flow from them, beyond preservation itself); Lauretta Burke, Elizabeth Selig, and Mark Spalding, REEFS AT RISK IN SOUTHEAST ASIA, 66–67 (World Resources Institute, 2002), available at: <http://www.wri.org/press/reefsat risk_bahasa.html>.

55. *Id.* at 66–70.

56. Historically, universal jurisdiction has been understood to apply to piracy, slave trading, war crimes, crimes against peace, crimes against humanity, genocide, and torture. See Princeton University Program in Law and Public Affairs, THE PRINCETON PRINCIPLES ON UNIVERSAL JURISDICTION, 28 (University of Minnesota Human Rights Library, 2001); Kenneth Randall, *Universal jurisdiction under international law*, 66 TEX. L. REV. 785, 815–39 (1988). Although universal jurisdiction developed long ago to provide a means of prosecuting miscreants (such as pirates or slave-traders) who otherwise might not fall under the jurisdiction of any nation, it has been extended within the past few decades by some nations to prosecute people who would likely escape justice in their home countries. Universal jurisdiction is used particularly against those who have allegedly committed crimes against humanity of sufficient magnitude to shock the global conscience, at least as viewed by the prosecuting nation. *See, e.g.*, Fiona McKay, Redress Trust Report, UNIVERSAL JURISDICTION IN EUROPE

(Redress, 1999) (describes efforts of victims of torture in other countries to seek redress in European state courts); International Law Association, Committee on International Human Rights Law and Practice, FINAL REPORT ON THE EXERCISE OF UNIVERSAL JURISDICTION IN RESPECT OF GROSS HUMAN RIGHTS OFFENCES, 3 (International Law Association, 2000). Universal jurisdiction has been aggressively applied by some nations such as Spain, which issued an international arrest warrant for the elderly former Chilean head of state Augusto Pinochet. On October 16, 1998, Pinochet was arrested in London by British authorities pursuant to this warrant, and a lengthy extradition controversy ensued. *See generally* Ruth Wedgwood, *Pinochet and international law*, 11 PACE INT'L L. REV. 287 (1999).

57. *See* Kunich, *supra* note 1, at 73–75, 169–70. Because my proposed statute would be limited to offering positive inducements rather than imposing punitive actions, it would avoid the problems attendant to Environmental Trade Measures (sanctions) identified by several major GATT/WTO decisions.

58. *See* Jeremy B.C. Jackson, et al., *Historical overfishing and the recent collapse of coastal ecosystems,* 293 SCIENCE 629, 635–36 (2001), available at: <http://geosci.uchicago.edu/Faculty/KIDWELL/Jackson2001Science Overfish.pdf>. *See also* Pauly, et al., *supra* note 33, at 689–91; Paul K. Dayton, et al., ECOLOGICAL EFFECTS OF FISHING IN MARINE ECOSYSTEMS OF THE UNITED STATES, 15 (Pew Oceans Commission, 2002), available at: <http://www.pewoceans.org/reports/POC_EcoEffcts_ Rep2.pdf> (hereinafter EFFECTS OF FISHING).

59. *Id.* at 635–36 (discusses the manner in which the many human-induced threats to marine biodiversity combine to generate a disturbance greater than any threat could cause in isolation). *See also* Marten Scheffer, et al., *Catastrophic shifts in ecosystems,* 413 NATURE 591–96 (2001), available at: <http://www.wau.nl/pers/01/scheffer-nature01.pdf> (describes how ecosystems such as coral reefs, oceans, and forests respond to gradual changes in climate, nutrients, habitat fragmentation, or exploitation with smooth change that can be interrupted by sudden drastic switches to a new and contrasting state).

60. As the philosopher Eric Hoffer remarked:

There are many who find a good alibi far more attractive than an achievement. For an achievement does not settle anything permanently. We still have to prove our worth anew each day: we have to prove that we are as good today as we were yesterday. But when we have a valid alibi for not achieving anything we are fixed, so to speak, for life. Moreover, when we have an alibi for not writing a book, painting a picture, and so on, we have an alibi for not writing the greatest book and not painting the greatest picture. Small wonder that the effort expended and the punishment endured in obtaining a good alibi often exceed the effort and grief requisite for the attainment of a most marked achievement.

See Eric Hoffer Quotes, available at: <http://www.phnet.fi/public/mamaa1/hoffer.htm>.

61. Kunich, *supra* note 1, at 177–83; John Charles Kunich, *Preserving the womb of the unknown species with hotspots legislation,* 52 HAST. L. J. 1149, 1243–50 (2001).

62. *See* Claudia E. Mills and James T. Carlton, *Rationale for a system of international reserves for the open ocean,* 12 CONSERVATION BIOLOGY 244, 246 (1998), available at: <http://faculty.washington.edu/cemills/ConsBiol 1998.pdf>; *Ocean blueprint, supra* note 20, at Ch. 29 (recommends some U.S. actions toward the furtherance of international ocean science).

63. *See* Richard W. Spinrad, *Do we know what we don't know?,* 12 OCEANOGRAPHY (3), 2 (1999) (mentions that not long ago we knew little or nothing about hydrothermal-vent bacteria, gelatinous zooplankton, and the coelacanth, and that these and other discoveries led to a profusion of new knowledge regarding chemosynthesis, evolutionary biology, organismal biogeochemistry, and ocean dynamics).

64. *See* Andrew Balmford, et al., *The worldwide costs of marine protected areas,* 101 PNAS 9694–97 (June 29, 2004), available at: <http://www.pnas .org/cgi/content/full/101/26/9694>. An estimated one million jobs would be created, not destroyed, by such an initiative. *See also* Larry B. Crowder and Ransom A. Myers, *A comprehensive study of the ecological impacts of the worldwide pelagic longline industry, report to the Pew Charitable Trusts,* 112 (2001), available at: <http://www.seaturtles.org/pdf/Pew_Longline_2002 .pdf> (explains that the highly damaging longline fisheries often actually lose money or are only marginally profitable while afflicting the world with their unconscionable amounts of by-catch).

65. Balmford, et al., *supra* note 64. *See also* Financing Protected Areas Task Force of the World Commission on Protected Areas (WCPA) of IUCN, in collaboration with the Economics Unit of IUCN, *Financing Protected Areas, IUCN* (2000), available at: <http://www.iucn.org/themes/wcpa/pubs/pdfs/Financing_PAs.pdf>; National Research Council, *Marine protected areas: Tools for sustaining ocean ecosystems,* 48, Table 4-1 (2001), available

at: <http://www.nap.edu/books/0309072867/html/48.html> (summarizes in tabular form the costs and benefits of a system of MPAs).

66. *See generally* ICRAN/Nature Conservancy/WCPA/WWF, *Marine protected areas: Benefits and costs for islands* (2005), available at: <http://www.panda.org/downloads/marine/50j185costbenefitsrap.pdf> (discusses the benefits to islands from MPAs in terms of more productive fishing industries, enhanced tourism, new jobs, and improved ecosystem services); WWF, *Marine protected areas: Providing a future for fish and people* (2005), available at: <http://www.panda.org/downloads/europe/marineprotectedareas.pdf>.

67. However, short of actual extinction, the number of individual members of some or many of the species in the hotspots may be significantly reduced in the absence of major preservation efforts. Over time, this diminution of population size could lead to reduced vigor, lessened genetic diversity, and greater vulnerability to disease, predation, or changed habitat conditions. In the long run, the extinction rate may be exacerbated due to our inaction, even without a current high extinction risk.

68. The enhanced protection of such hotspots could still provide a positive outcome in the form of greater viability of some of the species therein. Although most species would not have become extinct even without the heightened preservation efforts, the species would presumably benefit from more protection. They may enjoy an increase in population size, flourishing with more undisturbed habitat for breeding, feeding, and sheltering. This could eventually prove important in the event of an outbreak of disease, or devastating fire, floods, earthquakes, etc. A rise in numbers could supply a crucial cushion against future threats. Thus, the analogy to unused insurance is imperfect; even absent a major extinction threat, hotspots preservation can be expected to yield worthwhile benefits.

69. Blaise Pascal (1623–1662) was a brilliant French mathematician, scientist, and philosopher. His famous "wager" is one of the most intriguing of his many contributions. Simply put, Pascal's wager deals with our choice of whether to believe in God, or more accurately, our decision whether to believe in God and to live as if God cares how we live. Given that we cannot definitively determine God's existence or nonexistence nor discern the nature of God through objective, scientific means, what is the wise choice in light of the uncertainties? Pascal presupposed that God rewards belief and righteousness with eternal bliss and punishes disbelief and sinfulness with eternal anguish. Pascal posited that under these circumstances we should "bet" on God and live a righteous life; if we do, the rewards will be infinite for us if God exists, while our losses will be insignificant if there is no God.

If God exists and we reject God, we have lost everything, but if there is no God and we have believed in a fiction, at least we have led a good life and have not truly lost anything. Peter Kreeft, CHRISTIANITY FOR MODERN PAGANS: PASCAL'S PENSEES EDITED, OUTLINED, AND EXPLAINED, 292 (Ignatius Press, 1993).

70. Of course, there are more than two options. We could fund marine hotspots legislation at many different levels, and to a varying degree different spending levels may be adequate to protect some hotspots, or some portions of hotspots. Perhaps there would be a rough correlation between dollars spent and extent of preservation. But the underlying principles remain the same, and so for the sake of clarity we are considering only the two extreme options—large-scale funding, or none at all.

71. *See* Kunich, *supra* note 1, at 180–83.

72. *See* GESAMP no. 62, *supra* note 14, at 7–13.

73. *See* Callum M. Roberts, *Deep impact: The rising toll of fishing in the deep sea,* 17 TRENDS IN ECOLOGY AND EVOLUTION 242, 243 (May 2002).

74. *Id.* at 242–43. Larger vessels, more powerful winches, stronger cables, and rockhopper trawls have greatly expanded the reach of commercial fishing, and these enterprises have even been encouraged by government grants and subsidies (242). *See also* EFFECTS OF FISHING, *supra* note 58, at 27–29.

75. *See* Les Watling and Elliott A. Norse, *Disturbance of the seabed by mobile fishing gear: A comparison to forest clearcutting,* 12 CONSERVATION BIOLOGY 1180, 1191–94 (1998), available at: <http://www.stir.ac.uk/Departments/NaturalSciences/DBMS/coursenotes/28k7/bottom%20trawl.pdf>.

76. *Id.* at 1191–92. This astounding rate of destruction of one of the most vital marine habitats is almost impossible to imagine, but perhaps we can begin to grasp the magnitude of the wreckage when we equate the *annual* benthic losses to *all* of the area of the nations of Brazil, India, and the Congo combined, every year.

77. Adam Smith, AN INQUIRY INTO THE NATURE AND CAUSES OF THE WEALTH OF NATIONS (Edwin Canaan, ed., Modern Library, 1994).

CHAPTER FIVE

1. This famous utterance is attributed to the author Gertrude Stein, who was referring to Oakland, CA, where she spent her childhood. *See* Robert

Andrews, Mary Biggs, and Michael Seidel, eds., THE COLUMBIA WORLD OF QUOTATIONS (Columbia University Press, 1996), available at: <http://www.bartleby.com/66/37/55537.html>.

2. *Amistad* (Dreamworks, 1997).

3. U.S. Commission on Ocean Policy, *An ocean blueprint for the 21st century, final report, Washington, D.C.*, Ch. 25 and 29 (2004), available at: <http://www.oceancommission.gov/documents/full_color_rpt/000_ocean_full_report.pdf>. Chapter 25 makes recommendations for a "national strategy" for increasing scientific knowledge of the oceans. Chapter 29 advocates U.S. participation in the advancement of "international ocean science and policy."

4. Ronald K. O'Dor, *The unknown ocean: The baseline report of the Census of Marine Life Research Program*, 15, Fig. 10 (October 2003), available at: <http://www.coml.org/baseline/Baseline_Report_101603.pdf> (graphically depicts the estimated numbers of known and unknown species of the nine largest animal phyla, categorized by the oceanic realm in which they are found). *See generally* John Charles Kunich, *Losing Nemo: The mass extinction now threatening the world's ocean hotspots,* 30 COLUMBIA JOURNAL OF ENVIRONMENTAL LAW 1–133 (2005).

5. 16 U.S.C. 1531–1544.

6. 42 U.S.C. 9601–9675.

7. 33 U.S.C. 1251–1387.

8. 42 U.S.C. 7401–7671q.

9. 42 U.S.C. 6901–6992k.

10. 33 U.S.C. 2701–2761.

11. 42 U.S.C. 6992–6992g.

12. Peter Shaffer, AMADEUS: A PLAY BY PETER SHAFFER (Harper Perennial, 2001); *Amadeus*, directed by Milos Forman (Orion Pictures, 1984).

13. *Pinnochio* (Disney Films, 1940).

14. CNN, *Afghanistan's Taliban orders destruction of statues* (February 26, 2001), available at: <http://edition.cnn.com/2001/WORLD/asiapcf/central/02/26/taliban.statues/>.

15. *Camelot* (Warner Bros., 1967).

SUGGESTED READINGS

Jonathan E.M. Baillie, Craig Hilton-Taylor, and Simon N. Stuart, eds., IUCN RED LIST OF THREATENED SPECIES: A GLOBAL SPECIES ASSESSMENT (IUCN, The World Conservation Union, 2004).

Edith Brown Weiss and Harold K. Jacobsen, eds., ENGAGING COUNTRIES: STRENGTHENING COMPLIANCE WITH INTERNATIONAL ENVIRONMENTAL ACCORDS (MIT Press, 1998).

Lauretta Burke, Elizabeth Selig, and Mark Spalding, REEFS AT RISK IN SOUTHEAST ASIA (World Resources Institute, 2002).

Biliana Cicin-Sain and Robert W. Knecht, THE FUTURE OF U.S. OCEAN POLICY, 34 (Island Press, 2000).

Larry B. Crowder and Ransom A. Myers, *A Comprehensive Study of the Ecological Impacts of the Worldwide Pelagic Longline Industry, report to the Pew Charitable Trusts* (2001), available at: <http://www.seaturtles.org/pdf/Pew_Longline_2002.pdf>.

Paul K. Dayton, et al., ECOLOGICAL EFFECTS OF FISHING IN MARINE ECOSYSTEMS OF THE UNITED STATES (Pew Oceans Commission, 2002).

FAO, *The state of the world's fisheries and aquaculture 2004* (2004), available at: <http://www.fao.org/sof/sofia/index_en.htm>.

GESAMP (IMO/FAO/UNESCO/WMO/IAEA/UN/UNEP, Joint Group of Experts on the Scientific Aspects of Marine Environmental Protection), and Advisory Committee on Protection of the Sea, *Marine biodiversity: Patterns, threats, and conservation needs*. REP. STUD. GESAMP no. 62 (1997), available at: <http://gesamp.imo.org/no62/index.htm>.

GESAMP (IMO/FAO/UNESCO/WMO/IAEA/UN/UNEP, Joint Group of Experts on the Scientific Aspects of Marine Environmental Protection), and Advisory Committee on Protection of the Sea, *Protecting the Oceans from land-based activities: Land-based sources and activities affecting the quality and uses of the marine, coastal and associated freshwater environment*. REP. STUD. GESAMP no. 71 (2001), available at: <http://gesamp.imo.org/no71/index.htm>.

GESAMP (IMO/FAO/UNESCO/WMO/IAEA/UN/UNEP, Joint Group of Experts on the Scientific Aspects of Marine Environmental Protection), and Advisory Committee on Protection of the Sea, *A sea of troubles*. REP. STUD. GESAMP no. 70 (2001), available at: <http://gesamp.imo.org/no70/index.htm>.

Great Barrier Reef Marine Park Authority, The World Bank, and The World Conservation Union (Graeme Kelleher, Chris Bleakley, and Sue Wells, eds.), A GLOBAL REPRESENTATIVE SYSTEM OF MARINE PROTECTED AREAS, VOL. I (The World Bank, Washington, D.C., 1995).

Michel J. Kaiser and Sebastiann J. de Groot, eds., EFFECTS OF FISHING ON NON-TARGET SPECIES AND HABITATS (Blackwell Science, 2000).

Graeme Kelleher, GUIDELINES FOR MARINE PROTECTED AREAS, IUCN (IUCN Publications Services Unit, 1999).

R. Kramer, et al., eds., LAST STAND: PROTECTED AREAS & THE DEFENSE OF TROPICAL BIODIVERSITY (Oxford University Press, 1997).

John Charles Kunich, ARK OF THE BROKEN COVENANT: PROTECTING THE WORLD'S BIODIVERSITY HOTSPOTS (Praeger, 2003).

Simon Lyster, INTERNATIONAL WILDLIFE LAW: AN ANALYSIS OF INTERNATIONAL TREATIES CONCERNED WITH THE CONSERVATION OF WILDLIFE (Cambridge University Press, 1985).

J.R. McGoodwin, CRISIS IN THE WORLD'S FISHERIES: PEOPLE, PROBLEMS, AND POLICIES, 51 (Stanford University Press, 1990).

Norman Myers, ed., GAIA: AN ATLAS OF PLANET MANAGEMENT (Anchor, 1993).

National Research Council, EFFECTS OF TRAWLING AND DREDGING ON SEAFLOOR HABITAT (National Academy Press, 2002).

National Research Council, MARINE PROTECTED AREAS: TOOLS FOR SUSTAINING OCEAN (National Academy Press, 2001).

Ronald K. O'Dor, *The unknown ocean: The baseline report of the Census of Marine Life Research Program* (October 2003), available at: <http://www.coml.org/baseline/Baseline_Report_101603.pdf>.

Rupert F.G. Ormond, John D. Gage, and Martin V. Angel, eds., MARINE BIODIVERSITY: PATTERNS AND PROCESSES (Cambridge University Press, 1997).

Pew Oceans Commission, *America's living oceans: Charting a course for sea change* (May 2003), available at: <http://www.oceanconservancy.org/site/DocServer/oceans_report.pdf?docID=242>.

Pew Oceans Commission, *Marine reserves: A tool for ecosystem management and conservation* (2002), available at: <http://www.pewoceans.org/reports/pew_marine_reserves.pdf>.

O.T. Sandlund, P.J. Schei, and A. Viken, eds., THE SCALE AND ECOLOGICAL CONSEQUENCES OF BIOLOGICAL INVASIONS IN THE WORLD'S OCEANS (Kluwer Academic Publishers, 1999).

Philippe Sands, PRINCIPLES OF INTERNATIONAL ENVIRONMENTAL LAW I: FRAMEWORKS, STANDARDS AND IMPLEMENTATION (Manchester University Press, 1995).

Boyce Thorne-Miller, THE LIVING OCEAN, 2nd ed. (Island Press, 1999).

U.S. Commission on Ocean Policy, *An ocean blueprint for the 21st century, Final Report, Washington, D.C.* (2004), available at: <http://www.oceancommission.gov/documents/full_color_rpt/000_ocean_full_report.pdf>.

Cindy Lee Van Dover, THE ECOLOGY OF DEEP-SEA HYDROTHERMAL VENTS (Princeton University Press, 2000).

Jon Van Dyke, et al., eds., FREEDOM FOR THE SEAS IN THE 21st CENTURY, 274–76 (Island Press, 1993).

James C.F. Wang, HANDBOOK ON OCEAN POLITICS AND LAW (Greenwood Press, 1992).

Edward O. Wilson, THE DIVERSITY OF LIFE (Belknap, Harvard, 1992).

Colin Woodard, OCEAN'S END: A TRAVEL THROUGH ENDAN-
GERED SEAS, 43–44 (Basic Books, 2000).
World Wildlife Fund, *Marine reserves: Protecting the future of our oceans*,
available at: <www.wwfus.org/oceans/marine_reserves.pdf>.
WWF/IUCN, *The status of natural resources on the high seas* (2001), available
at: <http://www.ngo.grida.no/wwfneap/Publication/Submissions/
OSPAR2001/WWF_OSPAR01_HighSeasReport.pdf>.

INDEX

About the Author

JOHN CHARLES KUNICH is Associate Professor of Law, Appalachian School of Law, Virginia, and the author of several books, including *Ark of the Broken Covenant: Protecting the World's Biodiversity Hotspots* (Praeger, 2003).